The Gyroscope of Life

Understanding Balances (and Imbalances) in Nature

David Parrish

The Gyroscope of Life

Understanding Balances (and Imbalances) in Nature

Pocahontas Press
Blacksburg, Virginia

The Gyroscope of Life

Understanding Balances (and Imbalances) in Nature

Copyright © 2020 by David Parrish

All rights reserved. No part of this publication may be reproduced, stored in a retrieval system, or transmitted, in any form or by any means, electronic, mechanical, photocopying, recording, or otherwise, without the written prior permission of the author.

ISBN 978-0-9967744-7-5

Cover illustration by Christy Mackie
Author photograph by Joe Parrish
Book design by Michael Abraham

Printed in the United States of America

Pocahontas Press
www.pocahontaspress.com

The author can be reached by email at:
dparrish@vt.edu

To Carter, Tyler, and Evelyn, beloved grandchildren; and to Mona

Table of Contents

Dedication vi
Table of Contents vii
Preface viii
Acknowledgements x
1   Base Lines: Formative Years 2
2   Finding Home Plate: Normative Years 22
3   Naturalists, Biologists, and Metaphysicists 39
4   Creeds, Creations, and Chronologies 64
5   House Rules and Theatrics 92
6   Dead or Alive? And the Gyroscope of Life 112
7   Plant or Animal? 142
8   Male or Female? 172
9   Species, Evolution, and Domestication 201
10 Ecology, Ecosystems, and Agroecosystems 234
Further Reading 277
Author Bio 280

Preface

I am a part of all that I have met.

Alfred Tennyson (1809-1892)

This book is a love song to biology – a semi-nostalgic recounting of themes and principles learned in 50-plus years of biological study and practice. One goal is to show how the study of life can be surprisingly entertaining at many turns, while a second is to show why it is existentially important. We have the ability to make the study of life moot, and we sometimes seem hell-bent on doing just that. Mother Nature has three House Rules – three drop-dead requirements for a species' success; I will give my assessment of how well we are doing in complying. Spoiler alert: I have some concerns.

Here's another heads-up: I will be resorting regularly to what I call "autobiologic," bits of blue-sky musing and reasoning – not science as it is more deliberately practiced. These are occasions where, in dealing with some principle or phenomenon, I go off on explanatory or exploratory riffs, relying more on logic than on data. Many of the "autobiological" ideas I offer are not original with me. They spring from seeds planted by teachers, mentors, writers, students, grandchildren, etc. I am a part of all whom I have met.

To get us on the same sheet of music: The song begins with an account of where I am coming from, both literally and figuratively. The first couple of chapters, or we could call them verses, are autobiographical, covering my early life and continuing through my graduate training. By the time you finish them, I hope you will conclude that I have some biological creds, and I also hope you will see me less as an ivory-towered ex-professor and more like someone with whom you'd enjoy having a beer and talking about crazy, cool, and scary stuff. (Please feel free to drink a beer or two while you are reading. I consumed a few while I was writing.) Chapters/Verses 3 and 4 provide some of biology's historical, philosophical, and scientific underpinnings – points that make what follows less crazy and more cool, but still just as scary.

The overall tone of this love song is upbeat and informative, but be aware, a motif that appears early eventually becomes a major theme – that our species may be at a tipping point. Life has been very good to us, but it has also bestowed on us the power to destroy significant portions of life – to include the portion that is us. That is almost too heavy to include in a book that I want you to read. But, for the sake of those to whom this book is dedicated – and for everyone's grandchildren – please read on. I hope you come away with an appreciation for what life is and why it may be simultaneously both incredible and inevitable, but not inexplicable – nor inextinguishable.

David Parrish
Hole in the Woods
30 January 2020

Acknowledgements

Many people have participated in the birth of this book. The book's shortcomings are entirely on me, but I acknowledge here the very positive influences of others. My parents, Leon and Evelyn Parrish, instilled in me a love of learning and then let me go in new directions with the knowledge they encouraged me to obtain. Professors Frank Barclay, Ralph Amen, and Peter Davies were invaluable guides and mentors at critical forks in my professional path. Their specific roles are noted in Chapter 2, but suffice it to say here that this book would never have happened without their guidance.

Also in the category of making this book happen is my agent (and dear niece) Sharon Bowers. She urged me for several years to "write a book, write a book." I fended her off with protestations that nonscientists wouldn't want to read anything I might write. Hell, I wrote professionally for 30 years before I published anything that became widely cited. But in retirement, I realized that I wanted to write about biological matters for a nonscientific audience. Thank you, Sharon, for sowing that seed and for having faith that I could write this. In a similar vein, I acknowledge the book-birthing skills of Pocahontas Press. Jane and Michael Abraham were generous in their encouragement and in employing their expertise to turn a rather unruly project into what you are now reading.

I must also acknowledge and thank my beautiful, long-suffering wife. For 51 years, Janette has served as a sounding board for my ideas, and she has been periodically subjected to times where my writing and studies could have seriously infringed on domestic tranquility. This book has been another of those diversions. Yet, to her everlasting credit, she regularly listened to and suggested edits for passages under construction, and she served as a second set of ears as I read aloud the entire manuscript for a final edit.

And then, there is Mona. She will be further introduced – but not very flatteringly – in this book. (It is a complicated relationship.) However, I acknowledge her here for her inspiration and a lifetime of wonderment. I very immodestly think that she might approve of what I have written about her.

The Gyroscope of Life

Understanding Balances (and Imbalances) in Nature

1

Base Lines: Formative Years

Some people are born on third base and go through life thinking they hit a triple.

Barry Switzer (1937 –)

An apologia

Family lore has it that I was potty-trained when I was 1. If I was past diapers by the time I turned 1, that would have put me into a precocious category – at least in western cultures. But a child is technically still 1 at 23 months. Maybe I was that kind of a 1-year-old when I gained sphincter control. By current standards, toileting at 23 months would still be pretty early, but it would have been considered late 75 years ago. In the first half of the 20th century, being out of diapers by 18 months was the norm.

So, why my preoccupation with dirty diapers? I have none. In fact, I have never gotten past the stomach churning that goes with changing a really messy one. I broach the subject only to introduce a theme that will recur throughout this book; I am a product of my place and time. I see the world through lenses ground by others. My notions about everything from potty-training to gender, race, religion, and life itself are a product of the culture in which my life has been entrained.

Every good piece of writing tells a story. This book tells a story about biology. It will touch on biological history,

The Gyroscope of Life

philosophical underpinnings, and scientific principles. It will also offer some impressions that I have acquired in three-quarters of a century of participating in life, with most of that time spent studying living things. Because cultural entrainment molds such impressions, offering some details of my upbringing may provide insight into those insights. So, this and the next short chapter will be autobiographical but with some "autobiological" musings thrown in. I will be using "autobiologic," which is not a real word, in a mashup way, borrowing from the Greek for self (auto), life (bio), and reasoning (logic). Thus, autobiological ideas are derived from reason more than from data. As you will see, these ideas are often about matters for which data gathering is not possible. Autobiologic fits into the realm of speculation, hunches, or just musings.

A functional and geographical parsing of my life

My first 6 or 7 years were richly "formative." During that time, I developed from a fertilized egg into much of who I am today. In the next 20 or 25 "normative" years, I was engrossed in formal learning, with lots of informal education also thrown in. For the last 45 years, I have been a "practicing" biologist. Those three words – formative, normative, and practicing – are in quotes, because I am not using them in a precise, Websterian way. As applied to me, they mean what I want them to mean.

Those three functional life divisions – formative, normative, and practicing – correspond closely with the geographic divisions of my life. I spent my formative years (1943 to 1951) in East Tennessee's Appalachia. My normative years (1951 to 1975) found me initially still in Appalachia but

Base Lines: Formative Years

then in several other places across a couple of continents and engaged for much of that time in learning to be a biologist. I got my PhD ticket of admission to practice biology professionally in 1975, and shortly thereafter I relocated to Appalachia to profess and practice biology (and retire).

So, for historic and geographic reasons, I am going to dissect my life into three phases or stages. This chapter covers the 7-year formative stage. The next chapter will consider the 20-plus-year phase that I call normative. The rest of the book will occasionally mention formative and normative causes and effects, but its main focus will be on observations, insights, and musings of someone practicing biology. But, hey, we all practice biology, and then we die.

Some formative biological factors

In my first few years, the formative forces working on me clearly included genes that stamped me with blue eyes, little skin pigmentation, wavy hair, high arches, and perhaps a predisposition to knobby knees. Besides such easily identified physical attributes, my genes also influenced less readily quantified traits such as personality, cognitive ability, and health. With regards to personality, some autobiologic suggests that, for all of us, temperaments such as extroversion and empathy are at least partially a result of hardwiring and biochemical balances dictated by genes. I am an introvert, a trait that I am convinced is more closely tied to my genes than to my potty-training. Furthermore, I suspect such introversion genes could be an evolutionary vestige. Let's explore that bit of autobiologic.

If you've been around puppies much, you know one can see marked behavioral differences in litters. Some pups are

The Gyroscope of Life

more aggressive and growly. Others are more laid back and submissive. All are from the same parents and whelped into the same environment. Some of those personality differences, then, must have a genetic cause, and one can autobiologically reason that genes for various personality types might well have had survival value evolutionarily. In wild animals, the dominant and more aggressive individuals – the alphas, as they are sometimes called – might get to have more offspring and pass along their genes for aggression, but they also run some risk of being taken out of the gene pool before they get to reproduce. For example, their behavior may run them afoul of bigger alphas, either of their own kind or of another. A risk-averse, Type B individual might be more inclined to run away and live to make love another day – passing on its genes and improving the odds of its species' survival.

We can likewise surmise autobiologically that many human behaviors are instinctual or reflexive: hardwired and under genetic control. Infants leave the womb knowing how to cry. That sign of distress is not learned. Infants return smiles and initiate smiles before they can know the significance of a smile and seemingly without having to learn how to smile. In fact, they can charm us with smiles well before they have developed reasoning power. Likewise, lactating women don't learn how to "let down" the milk in their breasts when their babies nurse or sometimes just cry. That instinctual, reflexive behavior has biological value, but it can be problematic, or at least nonfunctional, if someone else's baby is crying.

But something besides autobiologic suggests that complex behaviors and personality traits are under genetic and hormonal control. We have data. Evidence comes from

Base Lines: Formative Years

studies with chimpanzees with whom we share 99% of our genes. The ape equivalent of a Myers-Briggs test reveals quite familiar-sounding personality traits: extraversion, conscientiousness, neuroticism, etc. More convincingly, neuroscience studies show such traits in chimps and in humans are under the influence of hormones. The hormone vasopressin affects feelings of aggression, love, and generosity, and levels of the hormone have been shown to be under genetic control. Numerous reports connect vasopressin (and oxytocin) levels in the brain with behavior and personality. Variations in vasopressin levels as well as in vasopressin receptors correlate with aggressiveness, empathy, attachment, etc. in humans, chimpanzees, bonobos, dogs, and even birds. So, I think it is likely that my introversion – shyness almost to the point of a phobia – is at least partially a genetic effect.

Some non-biological formative factors

Non-genetic influences undoubtedly also played a major role during my formative years. I absorbed many attitudes, beliefs, and behaviors not by training but by cultural entrainment. I absorbed osmotically the values, norms, and mores of the milieu in which I lived. This formative learning accrued into a set of assumptions about how the world operates. They were assumed to be true just because that's how things were in my world. To me, these truths were as constant, consistent, and integral as blue skies, setting suns, and shifting seasons. Of course, my "world" was small, and my perspective was limited and biased, but I didn't know that.

I know now that many of the cultural norms and societal patterns a 7-year-old, blue-eyed kid accepted as universal truths were representative only of the place and time into

The Gyroscope of Life

which I was born. And that kid had no idea how big the world was and no feel for the flow of time. As a 5- or 6-year-old, I had a particularly poor grasp of time; I thought we went to school for a year and then had vacation for a year. Summer was idyllic and lasted forever. Never mind that the "year" of vacation was always summer-like and the year of school included fall and winter. It was about how long it felt.

Religion was a formative force for me. Indeed, religion was a major force in much of society in those times – perhaps especially in Appalachia. Just about everybody I knew went to church somewhere and with some regularity. (I knew no one who went to synagogue or mosque.) Religion was an integral part of my early life. There was grace before every meal. My mother put us to bed with a Bible story followed by our now-I-lay-me-down-to-sleep prayers. Our family went to church every time the doors were open. That was at least three times a week. Two or three times a year, the church would host a "revival" with a visiting minister speaking nightly for up to two weeks. Some of those events would be held in a big, open-sided, canvas tent. Sweltering summer nights. Cicadas rattling the air. Bugs buzzing around bare incandescent bulbs. Hard folding chairs. Oil cloth posters pinned up, with the speaker's key points and requisite biblical citations neatly lettered and sometimes artfully illustrated. Hand-held (and hand-operated) fans, with their funeral home advertisements, stirring the steamy air. Squirming kids. "Amens" from those assembled. Raised-voice exhortations to be saved.

So I was, in essence and inevitably, born into a religion. My faith was more like an accretion, but I didn't know that. It was just the way things were. I certainly never questioned anything I learned in those years. The group our family wor-

Base Lines: Formative Years

shipped with prided itself in having a proper understanding of biblical truths. We would say, if pressed, that we were the only group going to Heaven. We were averse to alternative views or interpretations of the infallible Bible. At 6 or 7 years old, I was not about to challenge the party line. In fact, I totally toed the line.

Besides being raised to know that a triune God was something to be reckoned with, the three Parrish boys were also taught about a red-suited, bearded fellow who did some reckoning of his own. Santa Claus was real to me at least through my eighth Christmas, which came 2 days after my seventh birthday. But then I began to have doubts – especially when I began to realize Santa was polytheistic. A Santa could be found in every department store in town and also ringing bells at kettles on multiple corners. The Santa's Helper explanation didn't satisfy. My skepticism turned to agnosticism and then to outright denial of the Big Fellow's existence. I feigned belief till my ninth Christmas because I was not sure what the consequences of "Santatheism" might be for the gifts appearing under my stocking.

As a child growing up in the South in the late 1940s and early 1950s, segregation was never a subject of discussion – not in my family. In my very white hometown of Norris, Tennessee, segregation wasn't apparent to a 7-year-old. Likewise, patriarchy seemed normal and right. Feminists were not recruiting either gender in those days. Without any instruction, I had been instilled with a set of norms and mores that reflected the times and my birthplace. Neither the racism nor the sexism that I now know were present were problematic then. The skewness of any of the values acquired in my formative years became apparent and a moral issue only in my normative years and after. Life, society, and

The Gyroscope of Life

I moved on, but my core values remained fundamentalist, small-town, mid-20th-century Appalachian.

Such moral blind spots are problematic in the 21st century AD, but autobiologic reasoning suggests how accreted, unquestioned judgments could have had survival value in the 21st century BC. In a hunter-gatherer culture, there could be advantage in having a brain that automatically creates lists of "okay" or "not-okay" things – things with which others have registered comfort or discomfort. Information gained from unconscious incorporation of others' values or behaviors could save one's life. If an animal that suddenly appears is one that my family members have shown fear of, I might not stick around to find out why.

Or such accreted knowledge could be valuable in a potentially dangerous but novel situation. Consider a scenario where a prehistoric human is moving through a forest and comes upon a never-before-seen animal – one not in the list of okay or not-okay things. Does he or she have to analyze the risk associated with everything in this rapidly unfolding scene? No; anything in the scene that is already in a list of okay things can be ignored. No analytical thought processes need be involved. Only new, unfamiliar things must receive attention. Such accreted knowledge could have been quite valuable once upon a time. It is less so in a less hostile world.

Born on third base

Barry Switzer, author of the line cited under this chapter's title, is more famously known for his days as a successful football coach at collegiate and professional levels. It is interesting, then, that he drew his mini-parable from baseball. If he had drawn it from football, he might have said, "Some

Base Lines: Formative Years

people are born on the opponent's 25-yard line and go through life thinking they ran the kickoff back for 75 yards." He made a good choice; the pig iron version just doesn't have that pithy, metaphoric ring.

I was born on third base. And it was a triple that put me there – just not one that I hit. Rather I had hit the jackpot; a hat trick; a trifecta: I was born male, white, and American in an era when each of those attributes provided me major advantages. The genetic lottery landed me in maleness. My parents secured my whiteness. Then either God or more good luck put me in the USA and not in Tanzania or any of many other places where life is often a struggle.

How lucky was I? There were about 138 million Americans when I was born in late 1943. Of those, essentially half were male, and close to 90% were white. With an estimated 2.4 billion people on Earth at that time, the math reveals that I was the beneficiary of a triple birthright shared by only 2.6% of the planet. Now, add supportive, hard-working, lower-middle-class parents. And then throw in a modicum of intelligence and a generous portion of health. That is a recipe for success. In short, I was put into life's game not just on third base but well down the third-base line. Given that heritage and those times, my professional and personal success was almost guaranteed. It was my game to lose.

A Golden Era

My seven formative years were, seen in retrospect, a Golden Era. That time included the last year and a half of WW II: an interval marked by D Day, the Battle of the Bulge, revelations of the atrocities of Nazi concentration camps, brutal island fighting in the Pacific, and the dropping

The Gyroscope of Life

of atomic bombs on two Japanese cities. In many ways, it was the worst of times for those directly affected by the war. And, for those on the home front, it certainly wasn't the best of times either. Mothers and fathers lost sons and daughters. Wives and children lost husbands and fathers. Food and gas were rationed. Victory gardens were not just de rigueur, they were necessities. But I remember none of that. I was not quite a year and a half old on VJ Day. I was in the bosom of my family, relatively unaffected by and certainly unaware of the fact that the world was at war.

The Cold War began quickly after WW II ended. Lots of bomb shelters were built and provisioned, although not in my family. I'm pretty sure that I knew about nuclear bombs by the time I was in first grade. They were new and awesome and scary. I don't remember, though, being traumatized by the knowledge that I could be vaporized in an instant. That piece of information did not seem to make it into the part of my brain reserved for thunderstorms and things that go bump in the night.

All of my formative years were spent in Norris, which served as a company town for Tennessee Valley Authority personnel. My father went to work for TVA in 1933 – in the midst of the Great Depression – as one of hundreds who flocked to Norris to find employment building Norris Dam. That project gouged a foundation into bedrock and then formed a dam across the Clinch River to produce what is still the largest lake in the Tennessee River watershed.

My images of those idyllic formative years are undoubtedly skewed by the fallibility and malleability of memory. When I returned to visit Norris several years ago, after many years of seeing it only in my mind's eye, I realized how different one's perspective becomes when looking at the world

Base Lines: Formative Years

from two feet higher up. Those big houses I had lived in were actually quite small. The two miles that I walked to school turned out to be a half mile. But Norris was still amazingly familiar. In many ways, it seemed unchanged.

Norris had no traffic lights in those days and few street lights. It had a small business center, which included a grocery store, a bank, a drug store, a barbershop, a dry cleaner, the post office, town hall, and the fire station. To this day, US Mail is not delivered to Norris homes. Milk was delivered to the doorstep in those days, though. I remember looking out on the back porch one very cold morning in the late 1940s and being amazed how the cream, which always was on top in those days of non-homogenized milk, had pushed its way up and out of the milk bottle as the milk froze and expanded underneath it. I remember, too, watching Mom add packets of food coloring to the oleomargarine, or just "oleo", she brought home from the store. The standard explanation for this practice was that makers of the butter substitute, which had an insipid color, were not allowed to use colorants to make it look more butter-like. But, a colorant could be provided with the sale, and our family preferred spreading something on its toast that didn't look like lard.

All of the houses in Norris were designed and built by architects and craftsmen working for WPA or some similar federal agency. They were rented to TVA personnel and their families. Rental contracts were awarded to those at the top of a waiting list but also using some sort of priority or seniority system. That system faltered in 1948, when houses began to be sold into private hands. As a result of the shortage of rental places, our family of five spent the winter of 1949-50 in a log cabin at nearby Norris Lake State Park. Its cabins were meant for warm-weather use only. Ours had a fireplace

The Gyroscope of Life

along with bunkbeds, but that was about it. We were happy when my father's name got to the top of the waiting list for a rental back in town.

Norris is still semi-rural. It has developed on the edges, but the original neighborhoods are amazingly intact and still look much like they did in 1950. The homes our family lived in the 1940s and 1950s have been remodeled to the point that they no longer reflect my memories of them. But the place still retains the aura of welcoming openness in which I grew and thrived. Playgrounds and parks abounded and still do. Streets, almost all of which are called Something Road, still follow natural contours and are therefore amply curved. Most roads do not intersect at right angles. The half-mile route that I walked to school looks pretty much the same, although the school is no longer at the end of it. That route ran down an unpaved service road, through a neighborhood park, briefly down a tree lined road, along a shortcut through some woods, out onto a road winding its way up a natural dissection of the landscape, and then on up the hill to school. All of those roads, service roads, parks, and shortcuts are still there.

Perhaps because so many of the workers at the dam were young and married, there were lots of kids in Norris. This was also the beginning of the post-war baby boom. We lived in four different places during my time there. All, except the cabin at the Park, came with lots of playmates, and we played a lot. We could go wherever we wanted. The Parrish boys, though, were required to stay within earshot of our mother's whistle. She had an olive-drab, plastic, drill-sergeant's whistle that she would blow when it was time to come home. Its effective range was probably a quarter mile. So, we had a circle with a diameter of about a half mile that

Base Lines: Formative Years

we could claim as home territory. At one of our homes, that territory included the town's open dump – a place of endless discoveries.

One of my most vivid memories of Norris – and one that speaks volumes about the times – came from a winter day probably in early 1951. We had had a major snowfall, and because Norris hadn't plowed its streets, the snowy roads had become packed and icy. In those days, parents allowed kids to sled on the roads. Everybody knew kids would be out on some roads, and drivers, getting along just fine with their tire chains, drove cautiously. On this memorable day, the three Parrish brothers went up to the top of Oak Road, where it still ends in a cul-de-sac, and lined themselves up for a ride down the hill. At the crest of the slope, my older brother lay down on our steel-runner Flexible Flyer. My younger brother and I pushed the sled to get it going as fast as we could. Then I belly-flopped onto my older brother's back. My younger brother continued pushing and running alongside for another few paces, and then he flopped onto my back. And so, we were off! Not only did we get to the bottom of Oak Road, we still had enough momentum to turn onto Garden Road where we picked up a little speed, enough to carry us onto Orchard Road, and then onto Sawmill Road. We finally ran out of slope and speed, rolled off the sled, and began trudging back for another run. When I visited Norris with my younger brother a few years ago, we measured that sled run with his car's odometer: 0.6 miles. We may have set the record for the longest sled run on Norris public roads. If so, the record could still stand, because Norris now plows snow off of its roads, and even if it didn't, modern parents would not think of letting their children sled on them.

The Gyroscope of Life

Dross on a Golden Era

I'm not sure if Norris was segregated per se. It's not that it was homogenized racially. It was just homogeneous racially – all white. I'm sure black families lived in the area because a family photo shows a black woman who I was told helped with housekeeping and childcare after my older brother was born. But I knew of no homes where black families lived in this government-planned and -administered town. All of the faces in my first grade class were white. Old TVA photos of dam construction workers reveal no black faces. I don't remember seeing any, but I would suppose some black men and women worked in Norris. If they did, I'm sure they were subjected to the indignities of "colored-only" facilities. In the 1940s and 1950s, Norris was part of the Jim Crow South.

My first memory of blatant Jim Crowism came from perhaps about 1950. Our family was traveling to West Tennessee to visit aunts, uncles, and cousins in Dyer County. It's easy to get from Norris to Dyer County today. Just jump onto I-40, and you're there in about 6 hours. In the days before I-40, we would pack into the car before dawn and not arrive at our destination until after dark, even in the long days of summer. On one of these trips, we stopped at a restaurant in a small Middle Tennessee town. We were seated and perhaps already eating our meal when a black family came in – the parents and two or three children. They sat down at a table near the door. Someone from the restaurant came to their table and had an inaudible conversation with them. They got up and left. This puzzled me, and I asked my parents about it. One or both of them explained that the black family was not allowed to eat with white people. I don't recall that they editorialized. It seemed more like

Base Lines: Formative Years

a matter-of-fact reporting of what happened. I remember feeling that maybe it wasn't quite fair, but it was what it was. Thus, segregation was one of those formative societal norms that I accepted without scrutiny. Only in my normative years did I began to question the fairness and legitimacy of racial discrimination. At some point, it occurred to me that those black children had probably asked their parents the same question that I had asked mine. And when I became a father, it made my heart ache to think how difficult that answer must have been.

The mid-20th century was also a patriarchal time. This was the era of "Father Knows Best," which started on radio in 1949, and "Ozzie and Harriet," which first appeared in 1952 – shows that reinforced the message that men are leaders (and women should follow). My mother, who was bright and articulate, accepted her "lesser role" with grace. Indeed, I think she felt it was God-ordained and not to be gainsaid.

I have no recollection of hearing anything about homosexuality, and certainly nothing about LGBT as a 6- or 7-year-old. Hell, I hadn't even heard about sex, let alone its variations. I understood gender as a patently either-or thing, but I had no idea what the logic or biology was behind having both boys and girls. I just knew that God made us that way, and that was good enough for me. But I learned soon enough that all good Christians considered homosexuality perverse and an abomination. The oft-cited-but-probably-overstated 10% of the population who weren't straight must have struggled with their sexuality – especially those who were Christians.

I've come to realize that I nostalgically recall my formative years as gilded partially because of what they lacked. Some now-ubiquitous influences that were not around or that were

The Gyroscope of Life

easily avoided when I was a kid include digital games (to keep me inside), gratuitous violence portrayed as entertainment, private ownership of weapons designed (and used) to kill in large numbers, opioid epidemics, hate-based politics (McCarthyism aside), and so forth. All in all, during my 7 formative years, I grew up in a more kid- and people-friendly cultural environment – a veritable Golden Era. At least, it seemed that way then, and it feels even more so now.

The environment versus technology

While I didn't know it at the time, the natural environment generally was not so kid- and people-friendly. Norris provided its citizens a healthy place to live, but people living in areas that were more industrial and more populous often suffered from inattention to local environmental concerns. The technology that spurred industry was simultaneously spurring declines in air, water, and soil quality and creating quality-of-life issues for many. The Earth was being treated more like a disposal site than our home. In those days, fouling one's own nest was not just for the birds. In many places, untreated sewage and industrial wastes were spewed into streams and oceans. Particulates, gases, and volatiles from power plants, smelters, and factories were flung into the air with impunity. Solid wastes and chemicals were typically dumped or buried with minimal regard to the potential for hazardous materials that might leach into the soil and groundwater. As progressive and healthy as Norris was, that dump we played in was not a sanitary land fill. It was just an open space where garbage of any description could be dumped and burned. In that era, the environment was seen essentially as a bottomless pit. "Dilution is the solution to

Base Lines: Formative Years

pollution" was the operating principle.

Environmental stewardship did not have a prominent place in the national psyche in the first three-quarters of the 20th century. My father told me of hearing one of his neighbors in Dyer County, Tennessee boast, "me and my sons wore out three farms." This would likely have been in the 1920s, when farming operations in many places were still carried out by muscle and sinew. That section of West Tennessee is covered with wind-deposited soils. Such soils generally provide a wonderful medium for plant growth, but they are also highly subject to erosion by wind and water. With some fairly simple management techniques, the erosion can be greatly reduced. But this farmer and his sons considered the soil and the land to be expendable – something to use up and then move on. This same sequence happened on a wholesale scale – but perhaps with not as much bravado – in the 17th, 18th, and 19th centuries throughout the Appalachian region. Europeans would arrive, settle, clear forests, farm the rich topsoil that forests had formed, watch the topsoil wash away, try to farm on the subsoil that was left, fail mightily, give up (sometimes after multiple generations of trying to eke out a living), and move on. That same exhaustive, depletive mentality has characterized our use of other natural resources to include water, forests, wildlife, minerals, and fossil fuels.

In the first three-quarters of the 20th century, the general public seemed implicitly to believe in the wholesomeness of all technologies marketed to improve life. "Better living through chemistry" predates WW II, but it was still widely accepted into the 1950s and 1960s, as thousands of new chemicals were produced by the thousands of tons. I remember my father walking through his vegetable garden and

The Gyroscope of Life

dusting bean and potato leaves. That powder may well have been DDT. When we cleared out the home place after his death in 1991, we found an aged, half-full, 1-pound bag of 50% DDT – part of his store of several now-banned pesticides that were perfectly legal and acceptable for home use into the 1970s.

Medical science, part of the technological environment, was certainly well behind what it is today. Childhood diseases such as measles, chicken pox, and mumps were still common and without preventive vaccines. Polio or infantile paralysis was still a scourge. I contracted polio in the summer of 1948.

Lessons from a polio ward

The polio virus infects the sheath surrounding the neurons that innervate muscles. Without that insulating sheath, nerves cannot transmit their electrical impulses. The muscles become paralyzed. The day I went to the hospital in Knoxville (Norris had no hospitals), I had fallen and was unable to get up because the muscles in my shoulders and upper back were paralyzed. I was admitted to the hospital's polio ward, a room big enough to hold (as I recall 70 years later) eight or ten beds down each side of a central aisle. We were all children in this ward, although adults were susceptible to the disease also. Also in the ward were two iron lungs, large cylindrical machines into which some seriously ill patients were placed with only their heads sticking out. The machinery didn't just help them breathe; it caused them to breathe. These were children whose diaphragm and chest muscles were paralyzed. They could not draw a breath. The iron lung created a negative pressure within the cylinder, and their

Base Lines: Formative Years

chest would expand as air was pulled in through their noses and mouths. Some of those children got to go home in their iron lung, but many remained in it until it breathed their final breath.

Visitors, even family, were not allowed in the polio ward, which was on a second floor of the hospital. Every day my parents (and sometimes my brothers) came over from Norris, climbed a ladder that my father had made for this purpose, and talked to me through a closed window. During my hospitalization, my parents did not reveal their anxiety to me, but after becoming a parent and fretting over a child with a relatively routine childhood disease, I know that they must have been terrified. The outcomes of polio could include being crippled for life, being in an iron lung for a much shortened life, or death right there in the hospital. I was fortunate to have had a mild infection with only two weeks spent in the hospital. After a few months of physical therapy, essentially all traces of the disease were gone.

An incident that occurred while in the hospital became something of a flashbulb memory. Because the paralysis was in my back and shoulders, I was still quite mobile, as long as I didn't fall. One day, I sneaked out of the polio ward to do a little exploring. I walked around a couple of corners only to be confronted by a nurse in the corridor. She was a nurse from the polio ward who knew where I belonged. She said, "Why, you little stinker." It was probably said in the way you might suppose, with a smile and maybe even with a bit of admiration. However, I felt like I had been slapped. No one had ever called me a "stinker." I'm not sure I had ever heard the word, but I knew what stink and stinky meant. In my naïve little head, she had just called me something as bad as a dirty diaper. I was totally embarrassed and ashamed. She

The Gyroscope of Life

picked me up and took me back to the polio ward, but I smarted from that miscommunication for the rest of my stay in the ward – just another indication of how innocent the era, or I, was.

2

Finding Home Plate: Normative Years

It is utterly false and cruelly arbitrary to put all the play and learning into childhood, all the work into middle age, and all the regrets into old age.

Margaret Mead (1901 – 1978)

Stealing first base

While I was figuratively born on third base, I knew someone who literally stole first base. Vern, the father of a dear friend, was in the Navy during World War II. He was a pretty good baseball player, and his Navy team played in an exhibition game against legendary Leo Durocher's legendary Brooklyn Dodgers. Toward the end of one inning, with two out, Vern was up to bat. He had accumulated two strikes, and then he swung and missed. But the catcher dropped the ball. Perhaps because it was an exhibition game against amateurs, the catcher did not immediately pick up the ball and tag Vern, as was required to end the inning. Vern turned, started to saunter toward his dugout, and then glanced back to see that the catcher had tossed the ball back to the mound and was heading toward the Dodger dugout. He was off. He arrived safely at first base, and the inning continued.

I also once stole first base, but only figuratively. I will describe that later in this chapter. Lest I seem too proud of my feat, I must note here that Vern stole first literally and by guile. I managed it only metaphorically and only by sheer,

The Gyroscope of Life

dumb luck (with emphasis on the dumb).

Normative forces, neuroscience, and evolution

Using techniques with abbreviations like CT, PET, and fMRI, neuroscience has shown that the human brain may be primed to learn in a new way beginning around age 7. By that time, the brain has reached 95% of its adult size, and gray matter in the cerebral cortex has reached its greatest thickness. The prefrontal cortex, which is so important in executive function (planning, working memory, impulse control, etc.), emotion, and personality, has just finished a three-year spurt of growth. White matter formation, which began in the womb, will continue to age 25 or later. Presumably because of all these neurological changes, the brain becomes particularly good at storing, recalling, and utilizing information during the 15- to 20-year span that I have dubbed my normative years and that brain scientists describe as adolescence. (In brain science circles, adolescence extends well past the stage we usually describe with that word.) The structure of public education implicitly supports the notion that 6- or 7-year-olds are ready for a new phase in their training. We generally require a dozen years of formal education beginning at that point, and many of us voluntarily stretch the experience out for another 4 or 5 years. Some of us may remain in school long enough to use – and exceed – all of our adolescent brains' learning edge.

Autobiologic (my pseudo word for biological musings) suggests survival value could accrue to humans with brains that acquire information more readily as they move into adolescence. Prehistoric children might live safely in the bosoms of their families for a few years as their brains de-

Finding Home Plate: Normative Years

velop, but they would need to learn a critical set of tools before venturing far from the home fires. Among prehistoric humans, smart kids – those able to learn more quickly about people-eating predators, home territory, plants that look edible but are deadly, etc. – would be more likely to survive and therefore more likely to reproduce. As the children of early humans moved into their information-gathering adolescence, Mother Nature's single, no-nonsense Standard of Learning would have been: the child must be able to survive long enough to make offspring. Children better able to learn would be more likely to survive into their reproductive years and pass on their smarter genes.

Normative forces

Besides formal education, many other normative forces are at work during modern adolescence. These are ad lib and ad hoc as compared to official school curricula with their Standards of Learning. My informal normative learning came in an osmotic way from multiple sources: books, religion, peers, television, role models, and so forth. During this informal education, I came to perceive the world in a way that reflected those influences. Only later – perhaps as white matter made more and more connections between memory and reason – did I begin to understand how narrow that view was.

I would suppose this osmotic mode of normative learning became powerful at about age 7 largely because it was fueled by my brain's development, but it also coincided with when I began to read and when I first was thrown into more heterogeneous peer groups of kicking, biting, and farting kids. Until 7, my peer associations had been in our Norris

The Gyroscope of Life

neighborhoods and the church – a somewhat homogeneous grouping socioeconomically and behaviorally. When I went to public school after we moved to Johnson City, Tennessee, a somewhat bigger town than Norris, I encountered many different ideas and ways of doing things. I was 9 or 10 when our family got its first television. That opened a wholly new set of normative influences.

Thus, normative, as I am using the term in this chapter, has to do with what I learned through formal efforts of educators as well as lessons absorbed – more than learned – from being in the culture into which happenstance or good fortune plunked me. All this happened during a time when my brain was particularly adept at forming new patterns and connections. I became very much a product of what I learned formally at school and church and informally by peers, role models, and other influences. This chapter will highlight some of the normative forces and events that helped turn a happy 7-year-old kid into a happy 30-something biologist.

Public school years: Formal learning experiences

My first year of formal schooling in Norris is confirmed in my mind by few distinct memories. Everything back that far is pretty much a fog – vague impressions, with few specifics that can still be pulled out of the gray matter. My first grade classroom had a wall of windows that looked westward onto the playground. I have vague memories of participating in a maypole ceremony there, an Appalachian thing to do and a tradition borrowed from the European homes of our forebears. I know I learned the things first graders were supposed

Finding Home Plate: Normative Years

to learn, because I still have the report card from almost 70 years ago. (My parents were collectors and savers – almost hoarders.)

Second through fifth grades, now in Johnson City, were memorable at least to the extent that I can still recall my teachers' names. The schools I attended in the early 1950s typically set up their classrooms with wooden desks arranged in five or six rows of five or six desks each – all bolted to the floor. That style of desk had a seat that could be raised to get in and out, and the desk top often could be raised to store books and writing materials (including a bottle of fountain pen ink in the days before cheap ballpoint pens). The rooms had no open spaces for small groups to gather, as is the style in elementary classes today. Cork panels above the blackboards would have pictures to commemorate whatever holiday was approaching, or they might be festooned with letters – printed letters in first and second grades and cursive thereafter – or numbers or some other matter relevant to our studies.

Slate blackboards and chalk were the primary visual aid in 1950s classrooms. The blackboards were washed by the custodian before we arrived for class, so we started each day literally with a clean slate. After each lesson, all of the calculations or words on the board would be wiped off with felt erasers, which became increasingly less effective as the day went on. Getting to take the erasers outside to knock the chalk dust off was a special – and dusty – privilege. There were designated places – usually around the corner from street view – where it was okay to smack the erasers against the side of the brick building to dislodge the chalk dust. At some point, vacuum-powered eraser cleaners were introduced. Someone still got to do the honors, but it wasn't as much of an honor somehow, maybe because an unsupervised

The Gyroscope of Life

walk outside was no longer involved.

Holidays with religious connections were integrated into the elementary curriculum in those days. Observance of Christmas, Easter, and Thanksgiving included school closings, and the Christian connections were made clear in the classroom in the days leading up to the break. Valentine's Day was observed as well. Each student had a manila envelope, with his or her name on it, usually thumb-tacked to the wooden chalk tray running along the bottom of the blackboard. Each envelope would be filled with classmates' cards expressing Valentine wishes either in humorous or slightly sentimental ways. While such an exchange could have turned into a popularity contest, with some classmates getting few cards, that was not allowed to happen. It was made clear that everyone who wished to participate would put a card in each participant's manila envelope.

As I was about to enter sixth grade, our family moved into another elementary school district, and I ended up in a different school. The change of scenery seems to have stirred me socially. In fact, I became so social that my report card sported a C on Citizenship – also known as Deportment – at the end of one 6-week grading period. I really sweated taking home a C on my report card, especially in Citizenship. My parents were not going to be happy. So, in a panic, I erased the hand-entered C and replaced it with a B. Both entries were in ballpoint pen, and my erasure left much to be desired, but the forgery got me by at home. Phew! But then I had to sweat taking the report card back. I turned it in with my messily forged B still emblazoned on it. No blowback. I had stolen first base. I can only suppose the teacher did not notice the change and/or did not have a separate record of the grade she entered – or she was one hell of a sweet lady.

Finding Home Plate: Normative Years

Next came junior high, which predated today's middle schools and included seventh through ninth grades. I can't name many of the classes or teachers from junior high, but I do remember taking Latin and finally beginning to understand how – unlike English – a language might be put together in a systematic and sensible way. I also remember a ninth-grade general science course. It was a mixture of chemistry, physics, biology, and geology. I enjoyed it. I think it may have turned my attention to science. Our general science textbook had thought questions at the end of each chapter, and the teacher would sometimes use those to quiz us aloud. One of the questions went something like this: When a glass thermometer is put into a hot liquid, the mercury level may be observed to drop for a moment before it begins to rise. Why? It was a raise-your-hand question, rather than being asked to a squirming victim. I suspect I would have just squirmed if it had been directed at me, but, without that pressure, the answer came, and up went my hand. The answer: the glass tube of the thermometer will warm first, and it will expand. That expansion will increase the volume inside of the thermometer's bore and cause the mercury level to drop before it also begins to warm and expand. It was a small thing, but it stuck with me as you can see, and it gave me good feelings about science.

After junior high came high school. In the eleventh grade, I took biology along with third-year Latin, trigonometry, English, band, and junior ROTC. Biology was taught by one of the football coaches. He wasn't the world's best teacher, but I came out of Coach P's class knowing that I was going to be a biologist. So now I began to know generally where home plate lay for me. My college years were going to solidify and refine that choice.

The Gyroscope of Life

Informal learning outdoors

From even before my public school years, I loved being outdoors. Most of the kids I knew did too. The neighborhoods we lived in had lots of open space – forested and grassy expanses now largely developed into residential areas. But back in the day, they provided ball fields, engineering sites, sledding runs, firing ranges, kite launches, camping grounds, race tracks, swimming holes, fishing holes, and more. Lots of hours were spent in these areas playing, roaming, or just lolling around.

In my teen years, I got into hunting and fishing. The house we lived in was literally on the edge of town. Our property line marked the city limit. In season, I could step across the fence with my 16-gauge, single-barrel shotgun and hunt. In those days, the area around Johnson City had large populations of Bob White quail and mourning doves. I was never a good shot. Most of the quail and doves that I shot at kept right on flying. As time went on, I became less enthusiastic about hunting, but I still enjoyed target shooting. Fishing was the outdoor activity that I never grew tired of.

Spending time in the outdoors instilled some rudiments of biology. Poison ivy must be identified and avoided. Same for skunks. Poisonous snakes are especially to be avoided. Grape vines are good for swinging only for a year or so after they have been cut free from their roots. One also learns some practical biology from tree riding. Tree riding is an activity in which the rider climbs a small tree (with a trunk about the size of his leg) to near the top and then swings his legs wildly out into space while hanging on for dear life with his hands. If executed properly, the tree will bend and give an exhilarating ride to the ground. Letting go as one's feet

Finding Home Plate: Normative Years

near the ground is important, or the rider may be hoisted back to hang several feet up in the air. One also learns either through observation or personal catastrophe that poplar and pine saplings are poor choices for this activity. They can snap when the ride is only half over. Besides, pines are sappy, and that sap is hard to get off.

In those days, I was passionate about the outdoors and nature. (I still am.) In fact, I had a brief "poetic" period in high school years when I tried to convey that passion in verse. It was mercifully brief. One untitled piece of doggerel and teenage angst went like this: While bears brown> played near brooks clean,> you turned green to stone.> Don't cry me clear tears. And then there was a pseudo-haiku piece (although I'm sure I'd never heard of haiku when it was written) entitled "Necropolis": The Metropolis,> sad story,> pines for its rhyme's glory. As I said, it was a mercifully short period.

Some zeitgeists of my normative years

I did not understand until much later that significant portions of what I was learning in the 1950s and 1960s – my neurological adolescence and "normative years" – were time- and place-dependent. To some degree, I am still caught in the zeitgeists of my youth: trapped in time. My world view stems partially from information gained while my brain was primed for learning but not necessarily for discerning. I know I'm not alone. I suspect the inflexible nature of beliefs, norms, and mores accreted during adolescence has evolutionary value. Some autobiographical musing suggests such "set in stone" thinking could provide bonds with those of like mind, create communal solidarity, reinforce "us-ness",

The Gyroscope of Life

and help identify "them". That unexamined way of thinking would be good for building community and bonhomie within a band of hunter-gatherers. In modern times, it would seem to have minimal survival value. But it still seems to be readily tapped – especially by demagogues and cultists. Oops, have I wandered into politics and religion? I'll leave them right there. For now.

Public schools were segregated throughout my 12 years of enrollment. Schools in the South generally did not integrate until forced to by federal judges. The public school system in Johnson City began to integrate in 1961. But it did so one grade at a time, beginning with the first grade. At that point, I was a high school senior. So I never attended a public school with black classmates. The schools had black custodians and cafeteria workers, but there was little interaction with them.

The private church school I attended for my college freshman year was all-white, of course. Churches were essentially the last bastion of segregation in the South. In my sophomore year, now at a state school, I attended classes with black students. Granted, there were not many, but it was still a first for me. I had no epiphany, but my attitude about race began to change. As I met more black men and women in the classroom and later in the military, I realized that the racial divide was arbitrary. I now consider myself non-racist, but dark places in my mind still harbor shadows from that zeitgeist of my youth.

In my normative years, homosexuality was generally considered taboo. Both as a result of religious training and social stigma, I had a strong sense that male-and-female sex was the only acceptable form, with the further fundamentalist caveat that it was reserved entirely for those married. Well into my

Finding Home Plate: Normative Years

career, I finally came to understand that not only sexuality, but also gender is quite plastic biologically.

Sexism was in full display along with racism and homophobia, although I didn't realize it. In the 1950s, women were paid about 60% of what men were for the same week's work. The Equal Pay Act of 1963 began to correct that disparity, but limited upward mobility for women continued to mar the work place – to include in the sciences. The sexism inherent in fundamentalist Christianity, Judaism, and Islam might have been expected – if not excused – given the sacred writings of those religions. There were no such underpinnings to secular sexism. It was strictly a cultural bias.

Geopolitically, the US was clearly ascendant after WW II. It was the economic and military Super Power, but the Cold War quickly brought the threat of Mutually Assured Destruction. We probably had drills, but I don't remember practicing duck-and-cover, the euphemism for bending over as far as you can under your wooden school desk to kiss your ass goodbye. The Soviet challenge also extended into space. The launching of Sputnik I in 1957, when I was in the seventh grade, showed the US was behind USSR in at least one conspicuous area. That bit of hubris served to stimulate a renaissance in science education, perhaps at the expense of some other subjects, but it was the direction that geopolitics moved us.

College years

I started college in Fall 1962. That first year was at a small church-related school. I might have continued there, but something was calling me back to Johnson City – a young woman. I spent more time during my freshman year writing

The Gyroscope of Life

to her than I did in studies. So love beckoned, and I returned to Johnson City to enroll at East Tennessee State University as a biology major. ETSU was an educational bargain. Tuition was about $50 per quarter for as many credit hours as one might wish to take. Being a pretty decent student, I could take overloads, which in some quarters meant as many as 22 credits – seven or eight classes. I relished it, especially the biology classes. All of biology was interesting. There were so many different directions to go in biology. All I knew at that point was whatever direction I went would be zoological. Plants were for sissies.

But, then, I had to take the mandatory botany courses. In those days, every biology major at ETSU had to take three quarters of botany – BIOL 351, 2, and 3. Taking botany was perhaps a fig leaf to cover our zoological manliness. Frank Barclay taught all three quarters of botany. Frank was not the most inspiring teacher ever, but he inspired me. I don't know how, but before those three quarters were over, I was completely converted to botany. Looking back on our interactions more than 50 years later, I think it was Frank's unswerving love of the subject that made me think it would be okay to be a botanist. He took a genuine interest in his students, and he seemed to take pleasure in my budding botanical interests. Frank encouraged me to go to graduate school, and he was instrumental in helping me to identify Wake Forest University and to land a fellowship there.

Frank also got me interested in wild flower photography. On rainy days, he showed slides of plants we otherwise would have looked at on a field trip. He took great shots with his old single-lens-reflex. In 1967, I purchased a state-of-the-art single-lens-reflex, and I spent a lot of time in the field over the next 30 years looking for botanical subjects.

Finding Home Plate: Normative Years

Before digital photography came along to do away with Kodachrome film and Carousel projectors, I accumulated over 1,500 35-mm slides of wildflowers. Many of those early ones are amazingly vivid 50 years later. The total slide collection is well over 5,000 when family and travel photos are included. Anybody need some old Carousel trays?

At Wake Forest, I was fortunate to fall under the tutelage of someone who, again, coaxed the best out of me and instilled more into me. That was Ralph Amen. Ralph was a plant physiologist. He taught a graduate-level course in plant physiology that we all fondly called "plant philosophy." He was an excellent teacher – organized and thorough. He introduced us to thermodynamics, cybernetics, heuristics, and a lot of plant physiology. The lab exercises Ralph developed were designed to make us act like scientists. He was particularly painstaking in his grading of our submissions. When I got a lab report, term paper, or exam back from Ralph, I could be sure that whatever grade he put on it was fully justified. He knew of my tendency to procrastinate, and he took no pity. I did not get an A in plant physiology because of points lost for late submission of an otherwise good term paper. At Wake and under Ralph, I began to realize plant physiology wasn't just about taking a plant's pulse; it was about understanding how and even why plants have a pulse.

After a timeout from my biological education to meet my military obligation, I enrolled at Cornell in 1972 in a program that furthered my ambition to be a plant physiologist. For a third straight time, I gravitated to someone on the faculty with whom I clicked. Peter Davies, who was not much older than I, allowed me essentially free rein in choosing a research project, and then he benignly let me work my way through solving problems in plant physiology that I inevita-

The Gyroscope of Life

bly failed to see coming. It was a great learning opportunity, and it was when I began to think of myself as a scientist. I shall come back to that evolution in the next chapter.

Military years

The Selective Service Act of 1917 made it possible for the US to conscript male citizens for military service. While few might be called up in peace time, when the nation went to war, its young men were called on to provide muscle, bone, sinew, and blood. The rules changed over time, but by the time they pertained to me (1961), all males between 18 and 35 were required to be registered with the Selective Service. Then the Vietnam War made being drafted a real possibility. There were about 12,000 American Troops in Vietnam when I started college. That number had grown to almost 500,000 by the time I got my degree. The possibility of getting a college degree and taking it to Vietnam seemed to fill my horizon. I wasn't opposed to the war at that point, but if I went, I felt I could improve the odds of bringing that degree back home if I were an officer. So I enrolled in senior ROTC (Reserve Officers' Training Corps) at ETSU. That provided me a commission as a Second Lieutenant, Armor, in the US Army when I received my BS.

As it turned out, my first duty station was in Germany, and by the time that tour was over, the war in Vietnam was winding down. I had a good assignment in Germany. Although my commission was tied to tanks (hence the "Armor"), I spent most of my time as a staff officer, "flying a desk." But I did get to be on the tanks some. That included getting to be a Tank Commander (TC) during annual tank gunnery qualification. Our crew qualified as expert with

Finding Home Plate: Normative Years

me pretty much just riding along in our 50-ton M-60 tank, pointing the gunner towards the targets, and calling for what armament to be used. But there was one main-gun target that I was required to engage from my TC position. The target was identified by a puff of smoke coming from the hull of a tank about two miles away. I commanded, "Driver, stop!" and, "Loader, HEAT" (high explosive, anti-tank). Then I grabbed the control stick, wheeled the turret around to point the main gun generally at the target, dropped down into the cupola, got the target onto the range finder, dialed in the range, put the crosshairs on the target, announced "On the way," and pulled the trigger. Boom. A 20-pound, 4-inch-diameter HEAT round splashed onto that old tank hull. All in 10 or 15 seconds. What a rush! We did something similar at night, too.

    I did not put my botanical interests entirely on hold while I was in the Army. I carried my camera and a plant press with me sometimes when I went to the field. A plant press is a simple device for flattening collected plants between pieces of paper and cardboard while the plants dry. My plant press and plant collections drew a bit of ribbing from my fellow officers, but I was able to collect some interesting plants in the middle of German military training areas that would be off-limits to most botanists. My wife and I would also collect specimens while on trips to various parts of Europe. Several of those specimens are now in the herbarium at Virginia Tech.

Books in my life and vice versa

    Books were a big influence in my normative years. I absorbed subliminally and osmotically from my early read-

The Gyroscope of Life

ing days, when books were chosen just for the pleasure of reading stories. I absorbed attitudes, likes, and dislikes that are still with me – things like my fascination with the outdoors and old guns and my admiration for forthrightness. In my mid-normative years – high school and undergraduate days – I chose books with a view to being informed as well as entertained. In my later normative years, I went more for books with the clear intent of learning from them.

Beginning in the second grade and throughout elementary school, I read almost obsessively. During those years, I did much of my reading in bed after lights out. After Mom supervised the bedtime routine, she would turn off the lights and leave. As her footsteps retreated, I would turn on a dim light, prop a book against the headboard, and read while lying on my belly. (I'm pretty sure Mom was aware of the curfew violations, but she was a book lover, too.) The book was probably eight or ten inches from my nose. Young eyes are able to focus at that distance, but it's probably hard on them. I became near-sighted by the third grade and wore pretty thick glasses thereafter.

My love of books continued for many years, although, it was not as intense or obsessive as it was in those first few years. My interests branched out – still mostly fiction, but now with a bit more substance. In my college years, I fell in love with John Steinbeck and read everything of his that I could purchase in paperback. New paperbacks cost less than $1 in those days. After devouring Steinbeck and still in my undergraduate days (mid -1960s), I tried Erskine Caldwell. I read a half-dozen of his better known works. Both Caldwell's dismal portrayals and Steinbeck's dismissive portrayal of the South in *Travels with Charley* (1962) were normative forces in my re-evaluation – more likely, first evaluation – of my

Finding Home Plate: Normative Years

southern heritage.

I turned to non-fiction for a while in my undergraduate days – books like *For the New Intellectual* (1961), *Coming of Age in Samoa* (1928), *Psychology of Sex* (1933), and *Patterns of Culture* (1934). The non-fiction genre largely carried me through graduate school. The transactional analysis books by Eric Berne – *Games People Play* (1964) and *What Do You Say After You Say Hello?* (1972) – struck me intuitively as sound. I also read theological authors like J.B. Phillips – *Your God Is Too Small* (1953) and *Ring of Truth* (1967), both of which were affirmations of faith – and Albert Schweitzer, whose *Out of My Life and Thought* (1933) put living one's faith to the test. Other non-fiction books that got my attention and influenced my thinking in those times – and are still on my bookshelves – included *Silent Spring* (1962), *Sand County Almanac* (1966 reprinting), *The Naked Ape* (1967), *The Double Helix* (1968), *The Population Bomb* (1968), *The Peter Principle* (1969), *The Greening of America* (1970), *Future Shock* (1970), and *The Selfish Gene* (1976). Maybe I just liked short titles.

In sum, my normative years were a time for soaking up information and ideas about how the world works biologically and culturally. It was also a time for discovering and developing interests that have stayed with me well into my dotage. Much of who I am is simply a reflection of the time and places in which I grew up – cultural entrainment. Other influences were more structured and formal. Some of those notions, both accretions and those gained by formal instruction, have required revision as time goes on – a process that continues. This book has turned out to be a good exercise in that regard.

3

Naturalists, Biologists, and Metaphysicists

Flower in the crannied wall,
I pluck you out of the crannies,
I hold you here, root and all, in my hand,
Little flower—but if I could understand
What you are, root and all, and all in all,
I should know what God and man is.

Alfred Tennyson (1809 – 1892)

What's in a name?

I have known I would be a biologist ever since I took biology in high school in 1960-61. But, if I had been born 250 years ago, I could not have been a biologist. "Biology" and "biologist" appeared in the lexicon only in the early 19th century. Until then, people who studied plants and animals were called natural historians or naturalists, and their discipline was natural history. Those terms, or their Greek equivalents, go back to the time of Aristotle. Today, "natural history" seems archaic somehow, and I once supposed the phrase was going the way of words like betimes, damsel, and whilst. I was a bit chagrinned when it dawned on me that "natural history" is still in use and in some prominent places.

The third most frequently visited museum in the world – one that I have visited many times – is the National Museum of Natural History in Washington, D.C. Years before

Naturalists, Biologists, and Metaphysicists

I knew I would be a biologist, I was captivated by the huge elephants inside the main entrance, the high-ceilinged halls with reconstructions of dinosaur fossils and whale skeletons, the amazing insect collections, and the dioramas of primeval forests. I suspect my love of biology partially dates to those wide-eyed visits.

But one doesn't spend long in a museum of natural history — of which there are many — before realizing natural history is not just about living and once-living things. In the National Museum of Natural History are also halls and sections devoted to gems, minerals, earthquakes, and volcanoes. What was long called natural history is now more frequently described as biology and geology. The new nomenclature is not simply a modernization. Rather, it reflects a fundamentally different approach to the study of natural things. Natural history was and is about collecting, cataloguing, and curating life's and Earth's forms. Biology and geology are about using science to understand the hows and whys of life's and Earth's processes. This chapter will emphasize the evolution of natural history into biology, but a similar story can be told about how naturalists became geologists.

From its beginnings, natural history was more of an avocation than a profession. The field lent itself to amateur participation. Naturalists were often persons with education and wealth. They had time to spend in consideration of living things, gems, minerals, etc. In the 17th to 19th centuries, men of rank and leisure who amused themselves with such things were often called gentleman naturalists. But, in that era, some people began to study natural history professionally. Parson naturalists were ordained clerics who studied living things, rock strata, fossils, and minerals. During the mid-19th century, scriptural geologists served a similar role

The Gyroscope of Life

but solely in the geological area of natural history.

Plants, animals, rocks, minerals, gems, and volcanoes were easy enough for naturalists of any stripe to find, describe, and catalog. The main goal seemed to be acquisition and display of fascinating or hitherto unknown organisms and natural objects. It was not so much about studying nature as it was about curating curiosities of nature. Parson naturalists and scriptural geologists were more parochial – operating on the premise that a greater knowledge of creation would reflect greater glory on the creator.

To their everlasting credit, naturalists helped develop key concepts in biology. Charles Darwin, whose notions about evolution are foundational to biology, was considered a gentleman naturalist early in his career. According to some biographers, he contemplated becoming a parson naturalist before his mind turned to a consideration of the origin of species. The Cell Theory, which holds that all living things are made of cells, stems from the 17th century observations of Robert Hooke, who was an English architect as well as a gentleman naturalist and a "natural philosopher". Hooke, in fact, coined "cell" to provide the biological usage we give that word today.

Hooke's other gentlemanly field, that of natural philosophy, dealt with what we now describe as physics and chemistry. Aristotle, Galileo Galilei, and Isaac Newton are some of the more famous natural philosophers. No museum of natural philosophy can be found on the Mall in Washington, but departments of natural philosophy still occur in academia. In most places, though, they are just called departments of physics or chemistry.

As a biologist, I am kind of pissed that my profession's predecessors weren't also considered natural philosophers.

Naturalists, Biologists, and Metaphysicists

"Philosophy" is from the Greek for love of knowledge. I am sure natural historians were just as fond of the knowledge they gained as were the natural philosophers. I understand natural philosophers delved beyond readily observed things to reveal the nature of chemistry and physics. But, damn it, naturalists were philosophers too – not only in their love of knowledge, but also in trying to better understand the things they were cataloging and describing. Philosophical debates about life have gone on among naturalists for hundreds of years, and out of that history grew attempts at explaining – not just curating – life.

Philosophies about life and why or how it happens

When naturalists waxed philosophical, what did they have to say about life? What did they believe set living things apart from nonliving things? For two millennia, life was widely viewed by natural historians as the product of an animating force, or a vital energy – the philosophy of vitalism. Vitalists argued living things were alive because they had been imbued with a vital force. Life goes away when whatever is behind the curtain withdraws that animating factor. In human beings, the animating factor has often been equated with spirit, and some cultures still hold that spirits inhabit animals and plants as well. "Spirit" comes from the Latin for breath. This inspirational notion has biblical support. "The Lord God formed the man from the dust of the ground and breathed into his nostrils the breath of life, and the man became a living being" (Genesis 2:7). Chi (or qi for Scrabble buffs), which is a life force or energy flow in Chinese philosophy and medicine, is a Chinese word used

The Gyroscope of Life

otherwise for breath or air. Vitalism holds that life happens as long as an organism possesses the life force. If asked "why is there life?", a vitalist would say, "because it is granted to living things." For some naturalists – especially those moving toward becoming biologists – that answer did not satisfy.

About 400 years ago, some natural historians began to distance themselves philosophically from the vitalist explanation for life. These mechanists held that life occurs because organisms are put together in a specific, life-manifesting way. Life happens because matter is organized in such a way that it becomes a living mechanism. This was not a complete disavowal of any supernatural influence; many mechanists believed the precise organization needed for life could only be explained by some external force. If asked, "why is there life?" mechanists would say, "because organisms are put together in a life-producing order." That was a major step away from the vitalists' explanation, but it still failed to satisfy naturalists on the philosophical path to biology.

Reductionists are the modern version of mechanists. The school of reductionism gained wide appeal beginning in the 20th century as our understanding of biochemistry and biophysics grew. Reductionists are more curious about how life happens than why it happens. They are the kids who like to tear a toy apart to see how it works. They are not ready to concede that life is the product of a supernatural or undetectable force. They believe explanations for life lie behind the curtain, and they want to yank it aside. They believe strongly in causality – everything that happens is caused by something else that has happened. They feel that life is the complex outcome of a cascade of causes. Put another way, they believe the phenomena or behaviors that we call life can be broken down into a series of causes and effects.

Naturalists, Biologists, and Metaphysicists

Reductionists believe they can learn how life happens if they take living things apart and look carefully at how those parts work.

A reductionist's logic goes something like this: We will be able to understand what makes a cell a living thing if we can understand how its subcellular components (membranes, nuclei, mitochondria, chloroplasts, etc.) do what they do. To understand how subcellular components operate, we need to learn about the behavior of their constituents (proteins, lipids, nucleic acids, etc.). To know how those complex biochemicals interact, we need first to learn how simpler molecules interact. And so forth. It is a hierarchical approach: breaking down complex components into less and less complex layers and looking at those simpler levels to find causes for the effects seen at the higher levels.

Understanding how a car works can be tackled as a series of cause-and-effect steps. The car moves (effect) because its drive wheels are turning (cause). Those wheels are turning (now an effect) because of torque created by gears in a transaxle (cause). The transaxle is actuated by a spinning crankshaft. The crankshaft is spun by the sequential firing of pistons. The pistons' firing order is regulated by an electronic distributor and camshaft-operated valves. And so forth. This sequential cause-and-effect (reductionist) approach gives us some approximation of how a car operates. It's a hell of a lot better, anyway, than just saying, "it works because the engine is running," which would satisfy a vitalist. Only slightly better would be the mechanist's version: "it works because it was built in Detroit."

Francis Crick, co-winner of a Nobel Prize for revealing the nature of DNA, was a full-blown reductionist. He famously said in 1966, putting words in the mouths of many

The Gyroscope of Life

who may not have agreed with him, "The ultimate aim of the modern movement in biology is to explain all biology in terms of physics and chemistry." That is my scientific zeitgeist. That is the philosophical mold I was stamped in while in my normative years. I am a reductionist. Many biologists from that era consider themselves reductionists. We think reductionism can be used ultimately to explain life. And death.

The contemporary philosophical flipside of reductionism is holism, which is now often equated with "systems biology". Holism emerged within modern biology as something of a contrarian, or at least reactionary, view to reductionism. The holists' basic viewpoint, if I may put words in their mouths, is that an unpredictable synergy happens as we look at higher and higher levels of complexity. We cannot predict what will happen in complex systems when their component parts are considered in isolation because the isolated parts do not perform as they would in the intact system. Aristotle said a couple of millennia ago, "The whole is greater than the sum of the parts." That is the mantra of systems biologists. Life is not just a bunch of chemicals. It is something that occurs only because all those chemicals get together and interact in ways that we could not have predicted from studying them in isolation. And at each succeeding level of complexity, more and more of these unpredictable surprises emerge.

Scientific thought and philosophy are marked by ebb and flow. During the era when living things were in the domain of natural history, vitalism gradually gave way to mechanist thinking. As natural history evolved into becoming biology, mechanists' thinking evolved into reductionism. Reductionism was ascendant in the era when I was being trained. But, for many biologists now, both reductionism and holism have become pejorative terms. To some degree, the debate be-

Naturalists, Biologists, and Metaphysicists

tween reductionists and systems biologists is a pissing contest about whether biology is its own domain in science or just an application of physics and chemistry. It is fair to say that reductionist biology continues to run into some pretty knotty problems, ones that haven't lent themselves to break-it-down-to-figure-it-out kinds of analyses. A more balanced view suggests reductionism and holism are two sides of the same coin and each philosophy has value.

Natural history is born again, as a science

In the late 18th and early 19th century, "geology" and "biology" were coined to describe what natural history was evolving into. Until then, naturalists had been simply describing, naming, and cataloging things. Most naturalists believed the natural phenomena they inventoried were the products of a creator – of a creation. Consciously or unconsciously, that supposition probably kept them from getting beyond curating phenomena and into asking questions about how such phenomena occur. But natural historians and natural philosophers began to see patterns and processes that aroused their curiosities to the level of asking "why" and even "how." What causes this? How does this happen? Science was "invented" to answer such questions. Science is what makes biology more than natural history.

Science employs a rigorous logic, a strategically reasoned approach, for finding sound explanations for natural phenomena. That logic or approach is often described as the scientific method. The philosophical underpinnings of the scientific method are not familiar to most nonscientists, but everyone knows a related logic employed in criminal courts. In courts, the accused is presumed innocent until proven

The Gyroscope of Life

guilty. Guilt can be adjudged only if all reasonable doubt has been removed from jurors' minds. Something similar must happen when science is in a cause-seeking mode. When scientists suspect some cause for a phenomenon, they proceed logic-wise as though their suspect is innocent. To test its innocence, they perform an experiment in which the suspect can show it does not cause the phenomenon of which it is suspected. But, in science, failing the equivalent of a lie detector test does not prove guilt beyond all reasonable doubt – especially when other suspects are still out there – and most especially when the jury is made up of hundreds of scientific peers.

Culpability and innocence aside, let's look at how science moves us ahead in our understanding of the universe. Scientists observe a phenomenon and then make informed guesses or hypotheses about what may cause it. They might come up with a half-dozen reasonable guesses. Then they set out to experimentally prove each hypothesis is wrong. If only five can be disproved, however, the remaining one is not thereby proved to be the cause. It is clearly a good contender, but what about many more perhaps equally reasonable hypotheses that have not yet been considered?

People who do not understand the logic behind the scientific method are sometimes skeptical of reports expressing a 95% confidence in a scientific finding. Atmospheric and climate scientists are close to unanimous in thinking that global climate change is human-caused, but they will say, as they must, there is a chance that the climate changes observed are due to some as-yet-undiscovered factor. This provides some an opportunity to argue that more studies need to be done. The fact is, more studies will not prove anything to a 100% confidence level. That is not what science does. It

Naturalists, Biologists, and Metaphysicists

only eliminates possible explanations. If someone will come up with a hypothesis other than human actions for global climate change, scientists can try to disprove it, but they will never be able to prove it is the cause – only that it is a contender.

Neither humility nor reticence keeps a scientist from saying he or she has proven some hypothesis or theory beyond a doubt. Many scientists have egos that would allow them to make such a claim if they could, but they know they have only disproven some alternative hypotheses. It is simple logic. Scientists disprove every alternative explanation they can think of, but few are so arrogant to suppose that they have thought of – and ruled out – every alternative explanation. Beware of any scientist who does.

In fact, humility often goes along with good science. Scientists perhaps should be more aware than any other profession of just how ignorant we are. One of the most brilliant minds of the 20th century, Albert Einstein, said, "As our circle of knowledge expands, so does the circumference of darkness surrounding it." Herman O'Dell, one of my undergraduate mentors, was perhaps paraphrasing Einstein when he said with a twinkle in his eye, "The circumference of our ignorance increases 3.1416 times faster than the diameter of our knowledge." Natural philosopher Isaac Newton used a non-geometric metaphor when he said, "To myself I am only a child playing on the beach, while vast oceans of truth lie undiscovered before me."

So, scientists cannot prove explanations for phenomena, but through experimentation, they can disprove faulty explanations. They formulate explanations or hypotheses for what causes a phenomenon, and then they design an experiment to prove that could not be the explanation. If they cannot

The Gyroscope of Life

disprove it, it might be the right explanation. Let's talk that through. Let me tell you how I learned to "do" science.

A naturalist is born again, as a scientist

In actuality, even though I was a biology major, I was more a naturalist than a biologist for most of my college career. During that time, everything I knew about biology came from books, lectures, museums, field trips, and similar exposure. My mind was pretty good at cataloging and curating such information. I still know the old scientific name for mink. (It has changed in the last 55 years.) Although I'm a botanist, I still remember the names for a lot of bones and muscles. I was pretty good at collecting and cataloging information about living things – the same set of skills that naturalists have used for two millennia. But no science was going on.

My master's thesis about seed dormancy in a common weed didn't really employ science. I wanted to know why the seeds of this plant were unable to germinate when mature. Studies with many plant species suggest that any one of a dozen or so mechanisms could explain why a seed ultimately germinates when it does. I tested several of those mechanisms and then stumbled onto the right one when I left my seeds in a greenhouse with poor temperature control. I was simply an observer, a naturalist – not a scientist.

When I began my doctoral studies in 1972, plant scientists were interested in trying to understand what makes plant stems grow. In fact, the interest was more focused than that. Work of many botanists had shown that a plant hormone called auxin triggers the elongation of stems. What wasn't known was how auxin causes stems to elongate. My

Naturalists, Biologists, and Metaphysicists

ambitious dissertation project was to look at auxin-induced stem elongation and figure out what was going on.

To study biological phenomena, it is often helpful to have a model system. Guinea pigs have long been used as model systems in animal studies; they even gave their name to the use of surrogates for such experiments. Fruit flies are another model system from which we have learned – among other things – how genes give rise to eye color, body parts, and complex behaviors. The botanical equivalent of the fruit fly is *Arabidopsis*, a small European weed that can complete its life cycle in 6 weeks. It has been a veritable genie's lamp in revealing secrets of biology down to the gene level. I was working to unmask auxin before *Arabidopsis* became widely used as a Guinea pig. Instead, I chose to experiment with a time-honored model system – peas. Gregor Mendel, a 19th century Augustinian monk and the father of genetics, used "English" pea plants to reveal patterns in the inheritance of traits. So, I decided to work with pea plants.

Pea plants produce vines – technically, stems – that can be several feet long. They do a lot of growing, and plant scientists had already shown their growth is stimulated by auxin produced at the tip of the vine. For me, using whole pea vines to learn how auxin works would have been much too complex and messy, especially since they produce their own auxin. So, I cut short sections from just below the tip of pea stems, put them in a tube, and measured how fast they grew. Studying isolated stem pieces should help reveal how auxin makes them grow when still in place. (Let's hear it for reductionism!) Before I could use them as a model system, though, I had to learn how those pieces from a pea vine behaved.

Cornell set me up with a state-of-the-art device for mea-

The Gyroscope of Life

suring stem elongation (Figure 3.1). Such devices have been in use since the early 20th century. They are called auxanometers, from the Greek "auxain", growth. (Of course auxin also got its name from the same word.) At the heart of my "growth meter" was a device borrowed from engineers – an instrument that can detect a millionth of an inch of change. I could measure the elongation of stem segments (stacked in a tube and flushed with water) with such sensitivity that I could see the recorder needle move. In fact, I typically adjusted the auxanometer's sensitivity downward to keep the needle from going off-scale during an experiment.

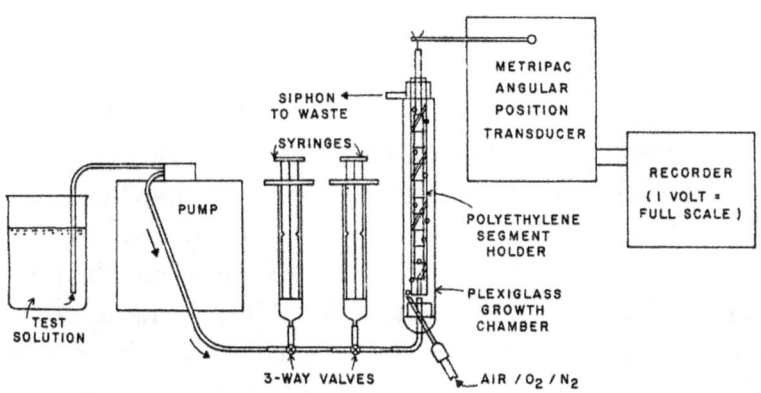

Figure 3.1. *The auxanometer set up. Short sections of pea stems are stacked in the segment holder, and solutions are pumped past them. The stems in the growth chamber can be well-aerated with air or oxygen (O_2), or they can be oxygen-starved by bubbling nitrogen (N_2). The transducer detects changes in the length of the stem pieces and sends a readout to the recorder. A typical readout is shown in Figure 3.2.*

Naturalists, Biologists, and Metaphysicists

I learned quickly that freshly cut pea stem segments grow rapidly when put into the auxanometer. More importantly to me, though, they did not respond to added auxin. However, by 4 or 5 hours after being cut, they would slow down to a lower rate. Most importantly for me, at that lower growth rate, they were responsive to auxin additions; within 10 minutes of adding off-the-shelf auxin, the segments would begin to grow at rates as high as those observed when freshly cut. Two possible explanations for these observations came to mind. The initially rapid growth could be due to the trauma of cutting and handling the segments, or it could be because the segments were still under the influence of native auxin that the stem tip had been sending to them. Or both. I didn't really care which at that point. I just wanted to avoid the 4- or 5-hour wait until segments were growing slowly and became responsive to added auxin. Those slower-growing, auxin-sensitive segments were to be my model system.

So, I decided to cut stem segments near the end of the day, hold them overnight in a beaker of water, and then do auxin studies with them the next day. After spending the night with no more trauma and away from the influence of auxin coming from the tip, they should be growing at a low rate and now responsive to applied auxin. Or so I reckoned. This was when things got really messy and when science finally started to kick in for me.

When I put 12-hour-old pea stem segments in the auxanometer, they sometimes behaved just as I wished, growing at a low rate and responding nicely to added auxin. But some days, when they went into the auxanometer, they took off growing like crazy. This was interesting, but it was also a pain in the ass. Repeatability is an absolute requirement in scientific research. Repeatability usually means someone else

The Gyroscope of Life

can do the same thing and get the same result. In this case, I couldn't repeat my own results! I was pulling my hair out. I spent the entire Christmas break of 1973 in the lab because I could not get the damned model system to behave consistently.

The breakthrough came when I made the observation (maybe the old naturalist instinct) that some stem segments were floating after their overnight in water, while others were lying on the bottom of the beaker. There were always some of each, but sometimes there were more of one or the other. (I still don't know why.) Up until this point, I had been using floaters and sinkers indiscriminately; stacking segments into the auxanometer without paying attention to whether they had been surface or bottom dwellers. I had implicitly been assuming each would behave the same once in the auxanometer. So it finally dawned on me to look at them separately. Eureka! The floaters behaved the way I wanted them to, while the sinkers were the crazy ones. I guessed that the sinkers were behaving so differently because they had become oxygen-deprived at the bottom of the beaker. Plant cells use oxygen, just as we do. In the bottom of an unstirred beaker, the oxygen in the stem segments' vicinity was perhaps getting used up, causing them to become, in essence, asphyxiated. Testing this hypothesis would be easy. Just bubble air in the beaker while it sat overnight. I did, and the issues with repeatability disappeared. The stem segments always came out growing at the desired low rate and responding within 10 minutes to added auxin. It seemed oxygen starvation was the bug in my model system.

Technically, I did not prove the sinkers were oxygen-deprived. Officially and scientifically, I tested what is called a null hypothesis: that aeration would not correct the prob-

Naturalists, Biologists, and Metaphysicists

lem. It did correct the problem, so I showed aeration does what I needed. But, although I am pretty sure, I cannot say with 100% certainty that it was the added oxygen that saved the day. Maybe, instead, it had something to do with the tumbling that inevitably happens in a bubbled beaker. Or maybe it had to do with a buildup of carbon dioxide or some other gas that is relieved by bubbling. I could design experiments to test those hypotheses, and the results would bring me a bit closer to certainty that the weirdness was all about oxygen deprivation. But all I cared about at that point was that my model system was now stable, and its results were reproducible.

But, at that same point, my research took a sharp turn. While I was pulling my hair out, trying to create a model system for studying how auxin works, I got scooped. Plant scientists at other institutions developed good evidence for a theory that auxin causes growth by stimulating cells to pump acid into their cell walls. The acidified cell walls become more plastic or stretchable as a result. This is the acid-growth theory for auxin's mode of action. It's still not proven beyond a shadow of doubt, but it's a damned fine theory. Which is to say, no one has shown it is not the cause of auxin-driven cell elongation. Plant scientists are probably 99% sure that this is how auxin causes stem elongation. In science, it doesn't get any better.

So, I had to go in a different direction for my dissertation work. Since it had slapped me in the face, I decided to look more closely at the effects of oxygen on stems' growth. I re-rigged the plumbing on the auxanometer to be able to alternately circulate aerated or oxygen-free solutions past stem segments while monitoring their elongation. Oxygen deprivation did what I expected; it shut down the growth of the segments within minutes. (Technically, I disproved a null hypothesis that oxygen starvation would have no effect on growth.) But, when I switched back to aerated solutions after anywhere from a few minutes to 2 or 3 hours without oxygen, the stem segments would take off growing at a rapid rate. They grew rapidly for so long that they more than made up for any growth lost during the oxygen-free interval – sometimes elongating 200 to 300% more (Figure 3.2). After 2 or 3 hours, they would slow back down to a low rate. I dubbed this previously unreported phenomenon emergent growth, using emergent in the general sense of a property that arises unexpectedly. I immediately determined I wanted to know how emergent growth happened – what caused it. And now I was beginning to develop some scientific chops to ask how.

My naturalist observational instincts were still at work. I noticed that stem segments undergoing emergent growth traced out growth curves that looked much like those of segments exposed to auxin. The timing and rates of response were strikingly similar – a 7- to 10-minute pause after restoration of oxygen or addition of auxin and then a sharp upturn. The new rate would be three or four times greater than that of stem segments bubbled with air overnight. Perhaps emergent growth was auxin-related? It made a reasonable hypothesis. I only had to develop some reasonable null hypotheses and then prove or disprove them experimentally. If I was dealing with an auxin-related phenomenon, it could provide an interesting addition to knowledge about auxin's *modus operandi* in regulating plant growth.

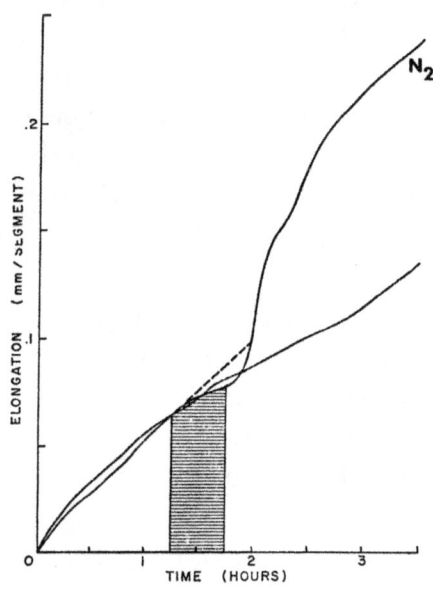

Figure 3.2. *An auxanometer readout demonstrating emergent growth. The lower, unlabeled curve shows the growth of pea stem pieces that are well-aerated for the full 3.5 hours of the experiment. The curve labeled "N_2" shows the greater growth of stem pieces deprived of oxygen for 30 minutes during the second hour of the experiment. The surprising difference between the two curves is a measure of "emergent growth".*

From long-standing work, we know the auxin that triggers stem elongation is produced in stem tips. The hormone is transported back to the region where auxin-responsive cells are located (typically about an inch below the tip), and it triggers cell elongation there. This auxin production, transport, and elongation mechanism is so sensitive and specific

The Gyroscope of Life

that plants use it to make their stems grow in one direction or another. A plant's stem grows toward light because auxin transport is favored on the shaded side of the stem. When the shaded side grows faster – because more auxin is arriving there – the stem bends toward the light. (If a plant is left in the dark or dim light so that all sides of the stem are shaded, the entire stem will grow faster and taller.) Something similar happens when a plant is turned on its side. Auxin transport is favored on the underside of the stem, and that side grows faster, causing the stem to turn upward. (Direction of root growth is also influenced by auxin transport and distribution, but it is the upper side of a horizontal root that is stimulated to grow faster.) So auxin transport and auxin distribution are a big deal for plants and for plant scientists.

I hypothesized that bubbling stem segments overnight allowed native auxin to be transported out of the segments. After being separated from the auxin-producing stem tip, time in the aerated solution would allow them to come down off of their natural auxin high. That seemed reasonable, since the segments aged overnight did slow to a low rate of growth that was quite responsive to auxin additions. But the emergent growth phenomenon suggested slow-growing stem segments might still have auxin in them, just not in forms or locations that could trigger rapid growth. Maybe asphyxiation released auxin and made it available to trigger rapid elongation – or so I hypothesized. That is a complex, non-testable, non-null hypothesis, however. Logic (and science) required that simpler, more-testable, null-type hypotheses be developed.

Long story short, I never proved emergent growth was caused by auxin that was sitting around in the stems, but I made a pretty good case for it. As part of the case, I let

Naturalists, Biologists, and Metaphysicists

stem segments take up radioactively labeled auxin and then measured loss of the radioactivity under well-aerated conditions. Loss of labeled material – presumably auxin – from the stem segments was rapid at first and then declined after 3 or 4 hours to a low, steady rate. I suspected that the rapid loss phase happened while the radioactive auxin was being pumped out of the stem by the auxin transport system. In parallel auxanometer studies, it took the same 3 to 4 hours to re-establish a steady, low rate of growth after an exposure to auxin. This made me think that it is the auxin in transport that is promoting growth, but I developed only circumstantial evidence at best – not valid in the court of null hypotheses.

But what would oxygen deprivation do to the release of labeled auxin? I hypothesized it would increase auxin efflux (flowing out). I hypothesized auxin was being held inside stems in a pool that is not closely tied to auxin transport or elongation. But that pool is presumably sensitive to a lack of oxygen, which would release the pooled auxin back into transport and thereby stimulate growth. Technically, I "null-hypothesized" that oxygen-free conditions would not increase the rate of radioactivity released. However, it did increase the rate of label leakage sharply, so the null hypothesis was disproven. But that is not the same as saying I proved oxygen deprivation causes native auxin to be released, which then causes stem elongation and emergent growth. Other hypotheses for emergent growth can be offered. For example, the lack of oxygen might temporarily increase the sensitivity of stem segments to low levels of auxin. Or oxygen deprivation might acidify cell walls without involving auxin at all. But I was at the end of my PhD work. The results were written up in my dissertation and four journal articles. I

The Gyroscope of Life

had demonstrated to my graduate committee and, with the journal articles, to the wider scientific community an ability to "do" science.

Science and religion

Over the centuries, debates among naturalists and biologists – whether between vitalists and mechanists or between reductionists and holists – have been generally tame and philosophical. By contrast, face-offs between theologians and scientists have tended to be contentious and personal. Such conflict might seem inevitable, because the stakes are high according to the lions and the pit bulls of both religion and science. But I shall make the case shortly that the friction stems from a great misunderstanding about what is at stake.

My life story captures in microcosm some portion of the religion-versus-science debate. I was raised believing in a trinity and a divinely inspired, inerrant Bible. In my youth, I was a fierce advocate for literal interpretation of the Bible, to include its creation accounts. I believed everything that now exists was made in 6, 24-hour days about 6,000 years ago. I scoffed at those who suggested the cosmos and life could be anything other than the handiwork of God. I was baptized at 12 to become a born-again Christian, and I continued toeing the fundamentalist line. (Interestingly, fundamentalism is a fairly recent movement in Christianity, appearing around the turn of the 20th century, partially in response to "Darwinism" and "modernism.") In my high school years, I read the Bible from cover to cover. When I went off to college, my freshman year was spent at a school where we had chapel every day and Bible classes were required every term. My faith was strong. Some biblical passages suggest such faith is

Naturalists, Biologists, and Metaphysicists

a gift. Some suggest it is a preordained gift. But I now look back on my early, unexamined faith as the natural product of a rather cloistered life. I was not exposed to much that might give me pause or cause me to consider alternative ways to view life and the universe.

But some of the facets of faith that had been drilled into me began to lose their luster as I increasingly rubbed up against the teachings and doctrines of other nominally Christian groups, usually in zealous mini-debates with chums. I say "nominally" because our sect, when pressed, would say it was the true church and only its adherents were going to Heaven. That sort of exclusivism is perhaps inevitable among fundamentalists, who maintain that the Bible is the inerrant word of God and that it alone provides a map to Him. The only way to reach God/Heaven/salvation was down the very narrow path my sect had identified from its studies of the scriptures. We were, as I eventually came to realize with great embarrassment, quite pharisaical.

During my later college years, I began to gain more direct knowledge of what science was saying about origins of the cosmos and life. Until then, my knowledge of such matters was gained entirely through the filtered logic and theology of Sunday school teachers and fundamentalist preachers. The more I learned about the diversity and adaptations of living things, the more I became persuaded that life likely had been going on for much, much longer than 6 millennia. I did not share such quibbles with fellow Christians. It took me some time to find a comfortable middle road – one that disconnects science and religion.

In spite of growing reservations, I continued to attend church faithfully. Faithfulness was pretty much measured by how many times a week one was present for services.

The Gyroscope of Life

Through my 50th year, our family was generally there three times a week, and we were frequent participants in and/or hosts for house-church activities as well. In the 1970s, I wrote a couple of articles for a now-defunct religious journal that identified with our sect but was decidedly liberal – an epithet not to be sought among fundamentalists. I remained a faithful, if heterodox, member of that group partially out of respect to my parents. My father died when I was 49. I stopped attending church the following year. So, I was a practicing – if latterly heterodox – evangelical for the first two-thirds of my life. I've been a practicing biologist – if initially a naturalist – for the last two-thirds of my life. I have lived the science versus religion debate, and I have resolved it in my mind into a nonissue.

The middle road

For some, my conversion-to-science story is a cautionary tale; this is what happens to those who choose not to believe. I would argue, however, that it was not a conversion, nor was it a choice, rather I came to understand science and religion are not inimical. They have no common moral high ground to fight for nor even to disagree about. They exist in separate, non-overlapping spheres – the physical and the metaphysical. The friction comes only when either sphere tries to push its way into the other's.

From where I sit on the bus, the science-versus-religion dispute stems largely from the opposing sides' disregard for the legitimacy of either faith-based/metaphysical arguments or fact-based/physical ones. Those who adduce the Bible or the Qur'an in support of their views do so quite legitimately in debates about the metaphysics of creation. But they do

Naturalists, Biologists, and Metaphysicists

themselves and the discussion a disservice when they use metaphysical sources to dismiss evidence for a natural (as opposed to supernatural) creation. The same must be said for iconoclasts who use scientific theories and findings in efforts to smash the beliefs of millions. In the end, science and religion are perhaps less like yin and yang and more like matter and antimatter. Each exists, but the two should not try to coexist. Each would do well to stay in its own lane.

Put into even more colloquial terms, science and religion should avoid mixed bathing. Mixed bathing is a term I grew up with. It described the gray area of unisex swimming. To avoid any issues, the Bible camps I attended each summer had separate swimming, a.k.a., bathing, times for boys and girls. There was no mixed bathing. Likewise, scientists and theologians should not be swimming together. Anyone who sets about to disprove the Bible, the Qur'an, the Sutras, the Vedas, etc. by adducing physical evidence is pissing in someone else's pool. Drawing on a deeply Appalachian colloquialism, anyone who uses sacred texts to dispute findings from science is barking up someone else's tree.

In defense of religions

Robert Heinlein, a well-known science fiction writer, said through one of his characters, "Religion is a crutch for people not strong enough to stand up to the unknown without help." No less a personage than professional-wrestler-cum-politician Jesse Ventura pretty much paraphrased Heinlein when he said, "Organized religion is a sham and a crutch for weak-minded people who need strength in numbers." Based on my experiences, I will not say religion is either a sham or a crutch. Religion offers much for its adherents – not the

The Gyroscope of Life

least of which is a sense of belonging and community. It codifies and encourages behaviors that are good for society at large. And it serves many as a source of comfort in the face of what might otherwise be overwhelming. Whether or not those good things are placebo effects is immaterial.

It was not inevitable that I become agnostic (or worse), given my scientific bent. Many excellent scientists are devout Christians, Muslims, and Hindus. Science and religion are not inherently at odds. Maybe my science-driven propensity to seek logic and consistency now makes it difficult for me to find resonance in religion because religion writ large offers a cacophony of dissonance. Maybe I am imposing my physical-world standards on the metaphysical world. I suspect some religions fall short of their aspirations. At least, that is my tentative conclusion when one religion or sect brands another as false or infidel. Certainly, I can offer no scientific proof regarding any religion. That would be a clear transgression of the metaphysical/physical divide. Besides, science isn't in the business of proving things…

4

Creeds, Creations, and Chronologies

In the beginning, God created the heavens and the earth.

Genesis 1:1

Biological apologetics and a biological credo

Life's diversity, complexity, ingenuity, and shear doggedness stir awe in us. The odd and fantastical measures that living things employ to protect, feed, and reproduce themselves are testaments to creative forces. Similarly, revelations about the vastness and freakiness of the heavens inspire amazement – even reverence. Out of such a sense of wonder perhaps inevitably grows a need for explanation. What are the causes behind life and the cosmos?

As discussed in the previous chapter, the search for answers to such questions tends to divide us into two camps. One group looks to science for answers, while the other generally invokes the supernatural. For simplicity's sake, let's call the two camps science and religion. I was raised in the religious camp. I understand the creationist argument, and I respect many of its adherents – to include some fine scientists. They stand on the premise that the universe and life are the products of a supernatural creator. It is one of their articles of faith. It removes the need for further explanations. We will look at a few creation accounts shortly. But first, let's look at the articles of faith that underlay the arguments of

The Gyroscope of Life

science – more specifically of biology.

"Apologetics" usually describes a systematic defense or argument explaining the rightness of a religion's doctrines and theology. Major belief systems (Christianity, Judaism, Islam, etc.) typically employ some version of a creed to lay out the fundamental tenets of their faith. Such tenets are established as foundational – not subject to question. Science is likewise built implicitly on a few fundamental assumptions – essentially biases that scientists adopt without real scrutiny. In what follows now, I shall offer an abbreviated apologia for biology, one that explains in credo fashion a logic leading to a peculiar view of and explanation for life. To understand how peculiar, almost heretical, that view is, consider this: many biologists feel a spontaneous development of life was inevitable. For many biologists, life is seen as an amazing – yet explicable – certainty; as unavoidable as death.

No formal biologists' credo has ever been drawn up to my knowledge. Table 4.1 is a personal creed, but I think many biologists would subscribe to it. It embodies relationships that we biologists see as laws of nature, as immutable in their own way as the law of gravity. It also includes some tenets that are more in the realm of articles of faith.

Table 4.1. A biologist's credo

I believe:
In causality: all objects, processes, and phenomena have natural causes;
Genes are an ultimate, highest-order cause of many biological phenomena;
Genes mutate and infrequently cause organisms to be better fit;

Creeds, Creations, and Chronologies

- Mutations that improve fitness will proliferate over time;
- This natural genetic shift causes organisms to differ from their ancestors;
- This evolution toward greater fitness is inevitable and produces new species;
- All living things are products of natural selection and evolution; and
- Natural processes produced the first life forms billions of years ago.

The first tenet – cause-and-effect, or causality – is a foundational, common-sense, first-principles concept in all the sciences. Every action, feature, or property of living and non-living things has a cause. Put the other way around, some cause gives rise to every observed effect. As a trivial example, I have knobby knees. Those knobs didn't just happen. They are the result of an inflammation that developed where the tendons from my kneecaps attach to my shinbones. The inflammation, likely a result of injury or over-use, caused a pronounced bump to form on the shinbone just below each knee. So, injury or overuse (a cause) led to inflammation (an effect). That inflammation (now as a cause) led to additional bony tissue being laid down (effect). As a generalization, biologists observe effects – biological phenomena – and then work backwards to try to find causes. In many cases, they find a falling-domino series of causes and effects. The causality sequence was short with my knobby knees, but it can be much longer for many phenomena.

As biologists work backward in a cause-and-effect chain to explain a biological phenomenon, they often find the ultimate cause is genetic. I am male because I have X and Y

The Gyroscope of Life

chromosomes, and the set of genes on those chromosomes directed formation of male components during my embryology and subsequent development. The belief in genes as a first-link-in-the-chain cause for many biological phenomena is, accordingly, part of a biologist's credo.

But we know genes are not a fixed, immutable cause. They turn out to be plastic, with a potential to mutate. Most mutations are undesirable or even fatal, but a precious few provide advantages – advantages that cause a lucky plant or animal to be better fit. That fitness is seen in the organism's increased ability to survive and make more of itself. Over time, that tendency – to produce more, better-fitting individuals – naturally causes the population of organisms to shift toward the mutant, better-adapted form. That sequence or process is natural selection.

Natural selection is a sacred tenet for biologists, but it can be a stumbling block for some coming from other fields. Maybe it will help to take the term out of its biological context for a moment. In the business world, natural selection works like this: Business A and Business B are selling gizmos in the same neighborhood, competing for the same set of customers. But Business A has a better business plan and is able to sell excellent gizmos at a lower price. It naturally attracts most of the gizmo sales in that neighborhood. Over time, Business A grows and opens new stores, and Business B goes belly up. Customer selection has "naturally" driven the entire sequence.

In the biological world, an organism's set of genes represents its business plan, its blueprint for making more of itself. Because of the mutability of genes, some organisms end up with a better business plan for a given environment. Those genes and the organisms that carry them will

Creeds, Creations, and Chronologies

be favored by natural selection in that environment. The gene flow follows the path of least resistance – or greatest fitness. We can understand, then, why a species' gene pool will change over time. In fact, after many, many generations, the changes can be so great that the organisms no longer look like their ancestors. That is the supreme autobiologic (remember: musings that aren't always data-driven) behind Charles Darwin's magnificent contribution to biology – evolution. More will be said about evolution both as a cause and as an effect in later chapters. Biologists consider evolution a logical and inevitable fact of life – inferable in fossil history and demonstrable in real time. And they argue that, if evolution didn't happen, it should. It is a natural consequence of introducing genetic variation into the reproductive process.

Somewhat parenthetically, and not an item in the credo: William of Ockham was a 14th century theologian and philosopher who suggested that, when solving problems, the solution that requires fewer assumptions is more likely to be true. This bit of logic is sometimes called the Law of Parsimony. More frequently, it is called Ockham's Razor. Ockham's Razor is often invoked in theological, mathematical, and scientific problem solving. Einstein used it in developing special relativity. We all know that the least complicated solution for a problem is not always the correct one, but Ockham's Razor still provides a good rule of thumb.

Belief in cause-and-effect, mutable genes, survival value, fitness, natural selection, and evolution leads to an Ockham's Razor notion – a bit of autobiologic – that life is a no-frills phenomenon. Everything about living things has a purpose. Every form, every action, every detail has a logical, functional explanation. Humans have vestigial parts with no current purpose, e.g., tail bones and appendices, but those are evolu-

The Gyroscope of Life

tionary remnants of once-useful parts. Living things simply cannot waste energy on producing doodads and gewgaws. Energy must be conserved for growth, resource gathering, protection, reproduction, etc. Evolution is parsimonious; newly evolved features improve fitness. Accordingly, biologists generally do not subscribe to the notion that leaves turn beautiful in the fall to inspire us, nor that the dazzling iridescent patterns of some butterflies and birds are meant for our enjoyment. Those phenomena are not frills. Biologists take it as a given that they have survival value – that they increase fitness.

Most of the items in the biologist's creed are established biological facts – not just articles of faith. We know that a set of genes is the underlying cause for much of what happens in an organism's life. We know those genes are mutable, and mutations lead to variation in gene sets. We know that genetic variations occasionally improve the fitness of an organism, and we understand the logic of why those with better business plans are naturally selected. We know that, even over the course of a few years, natural selection can produce organisms that look and behave differently than their progenitors. These are facts recognized by those inside and outside of biology.

But the first and last articles of the credo do not spring from factual knowledge or observation. They are closer to being articles of faith. They are ultimately unprovable and therefore require some measure of blind acceptance. Consider the causality tenet, for example. Scientists cannot disprove the possibility of supernatural causes, but they choose to play science by rules that say every effect has a natural cause. Hence the first article in the creed. If you will, call it a bias. Call it a misapplication of Ockham's Razor. It is an article of

Creeds, Creations, and Chronologies

faith.

Nor can science prove that life is billions of years old. Earth offers evidence that it formed over 4 billion years ago, and rocks that appear close to 4 billion years old show signs of ancient life. Likewise, astronomical observations and some simple calculations suggest the universe is close to 13.8 billion years old. But science cannot rule out the possibility of supernatural explanations, ones in which the universe and life were created in the last 6,000 years but made to look much older. Again, scientists tend to instinctively trust Ockham's Razor – and their causation prejudice – that the universe and life are just what they appear to be, products that have been billions of years in the making. Hence, the last two tenets of the credo.

Some creation accounts

Most cultures have developed their own creation accounts. Hundreds of culture-specific genesis stories have been cataloged. The accounts vary widely, but most involve supernatural beings. One straightforward creation story from the eastern Mediterranean suggests time began about 6 millennia ago, and everything in the universe was created in the next 6 days by the Elohim – subsequently known as Yahweh, or simply God. From an African tradition comes the story of Roog, a deity who formed the universe, Earth, and all its beings over an extended and elaborate sequence. A creation account from ancient China suggests the universe burst into existence from an egg-shaped cloud in which had developed Pangu, an Atlas-like giant. When Pangu died many millennia later, his body gave rise to the Sun, the Moon, and Earth as we know it today. Among Native Americans, many cre-

The Gyroscope of Life

ation stories cite some demiurge who brought the world – or at least a particular tribe or nation – into being.

Such metaphysical explanations for the universe and life – especially those with a divinity who has vested interest in one's own culture – have a powerful pull. They provide solace, sense, and salve against a baneful, bewildering, and bruising world. And, as noted in the last chapter, science cannot disprove such accounts. Metaphysics is outside the pale of the physical, or natural, sciences. Using science to critique cultural traditions rooted in mysticism or religion is like using cricket rules to referee a baseball game.

To be sure, scientists also want explanations for what they experience. However, their causation bias tends to push them toward seeking natural causes. Accordingly, science has developed its own creation account, using findings from astrophysics, geology, paleontology, radioisotope dating, biology, etc. Science now offers a sequence of events that explains, in a cause-and-effect way, the existence of the universe, Earth, and life. Science-based clues about the origins of the universe and life practically self-assemble into a creation story, one that requires neither belief nor disbelief in deities. In fairness, some unknowns are embedded in science's creation theories – areas where we brush against the limits of current knowledge. Such fuzzy areas require faith in the doctrine of causality and may deserve healthy skepticism.

From The Big Bang to lots of big bangs: Cosmology in a nutshell

Since about 1930, astronomers have known that the universe is expanding, and it appears to have been doing so for a long time. A back-calculation of the current rate of

Creeds, Creations, and Chronologies

expansion and other observations suggest that the expansion started from a single point around 13.8 billion years ago in an event named – with whimsical understatement – The Big Bang. "The Beginning" might seem more appropriate, but that phrase is perhaps covered by divine copyright. Key notions behind The Big Bang theory came from a Catholic priest who was also an astronomer. Ironically, because of the theory's ecclesiastic origins, there was some early resistance to it among astronomers – until the evidence became overwhelming.

Cosmologists' creation story and chronology go something like this (Table 4.2). Within the first trillionth of a second after The Big Bang, quarks, the seemingly irreducible elementary particles of matter, began forming. Energy was being converted into matter, just as Einstein suggested could happen in his momentous $E = mc^2$. By one millionth of a second after time started, quarks began to coalesce into subatomic particles, such as protons and neutrons. By about one second into the universe's life, essentially all subatomic particles now in existence had formed. The protons and neutrons quickly assembled themselves into atomic nuclei, but electrons did not join them to form non-charged elements for another 380,000 years. Hydrogen, the smallest, simplest element, was the most frequent product of this gestational period, but some helium and just a bit of lithium was formed also.

The story of how the cosmos came to have components like stars, galaxies, and galaxy clusters is much longer. A highlight reel would show a few key points (Table 4.2). The first stars likely began to ignite 50 to 200 million years after The Big Bang in a star-birthing process still going on all over the cosmos. Stars by the billions began to gravitationally

The Gyroscope of Life

huddle together into galaxies when the universe was only a few hundred million years old. Then groups of hundreds to thousands of galaxies became gravitationally tethered into galaxy clusters. Then galaxy clusters formed into superclusters, which were once considered the largest known structures in the universe. That distinction is now a matter of debate among astrophysicists and cosmologists. The age of the oldest galaxy clusters and superclusters is another point of contention as well. (As we will make clear shortly, biologists also quibble over their science.)

Table 4.2. The likely sequence in development of increasingly larger configurations of matter in the universe. Times in parentheses are dated from the instant of The Big Bang.

Quarks, electrons, and neutrinos form (0.000000000001 second)
↓
Quarks coalesce into protons and neutrons (0.000001 to 1 second)
↓
Naked (without electrons) atomic nuclei form (1 to 15 minutes)
↓
Electrons join nuclei to form simple atoms (380,000 years)
↓
First stars ignite (50 to 200 million years)
↓
Galaxies begin forming (200 to 400 million years)
↓
Galaxy clusters begin forming (1 to 3 billion years)
↓
Galaxy superclusters begin forming (2 to 5 billion years)

The cosmological sequence suggested in Table 4.2 would have produced a universe full of three simple elements,

Creeds, Creations, and Chronologies

trillions of stars, and a billion or more galaxies. But totally missing are all the other elements and minerals needed to make rocks, tectonic plates, and continents – not to mention life. A couple dozen elements are needed to make living matter, and then there are silicon, aluminum, gold, uranium, etc. – all the other elements in the periodic table that make up Earth. Astrophysicists' story of when and how those elements formed requires a closer look at the life of stars.

Stars have life cycles typically measured in billions of years. They begin as gravitationally condensed clouds of hydrogen, and under the extreme pressures and temperatures that develop at the cloud's core, hydrogen atoms begin to melt together. That is, hydrogen begins to undergo nuclear fusion – the physics in hydrogen bombs – to form helium. In the process, a star is born. Eventually it will die, though. Generally, a star less than one-third the size of our Sun will exhaust its hydrogen and die rather quietly, but it may take up to a trillion years to do so. (The universe is only 0.0138 trillion years old, so those smaller stars are going to be around a long, long time.)

The Sun and other stars of its size and a bit larger put on more of a show. As it starts to run out of hydrogen – about 5 billion years from now – the Sun will begin to fuse helium into carbon and oxygen. It will swell up into a red giant, but it will eventually burn though all its fusible material, cool down, and shrink into a white dwarf. At that point, the solar system will pretty much have run its course, although the planets and other gravitationally connected objects will continue to spin around the remnant of the Sun for eons and eons.

Stars at least eight times more massive than the Sun meet much more spectacular deaths. After they burn through their

The Gyroscope of Life

hydrogen and then their helium to make carbon and oxygen, they heat up enough to fuse carbon. Fusion of carbon produces larger, heavier elements such as iron, silicon, sulfur, and magnesium. But the element-forming process is about to take a spectacular turn. The star's core accumulates iron and begins to collapse under gravitational forces. The supercompressed core eventually heats to billions of degrees and suddenly annihilates itself in a titanic explosion – a supernova. In the cataclysm, the supernova creates and hurls into space many of the more complex elements needed to form life and rocky planets. Depending on its initial size, the supernova remnant will shrink back into a neutron star or a black hole.

Astrophysicists had postulated that there could be even bigger celestial shows than those produced by supernovae. While I was writing this book, their predictions were supported, as two neutron stars – supernovae remnants – were observed colliding and merging with one another some 130 million light years away from Earth. This was the first time such a kilonova event had been observed and data collected from it. The forces, radiations, and elements produced by the neutron stars' melding exceeded that of a supernova by far. Finger prints of essentially all the elements of the periodic table were observed in the glow of the blast. The amount of gold produced was estimated to be about ten times the mass of Earth! Talk about some serious alchemy. Also produced in massive quantities were lead, uranium, and everything else in the periodic table.

So, cosmologists and astrophysicists can offer a sequence of events – let's call them causes and effects – to get from a universe made only of energy to one that has produced all the known elements. In super-colliders, they can observe

Creeds, Creations, and Chronologies

some of the phenomena that characterized the first seconds of The Big Bang. Furthermore, they can observe and measure star formation, supernovas, and kilonovas and show that their theoretical predictions match up well with reality.

Making the third rock from the Sun

Evidence suggests that the Earth began forming about 4.6 billion years ago (bya), or about 9 billion years after The Big Bang (Table 4.3). This was after millions of stars had died gloriously and spawned the elements of which the Earth was to be made. The Sun was getting its act together at the same time, and it was holding lots of space junk within its gravitational sphere of influence – the growing solar system. Earth was initially just a cosmic dust ball, a gravitational accretion of shrapnel from element-forming supernovas and kilonovas.

More cosmic dust balls, besides nascent Earth, had coalesced in the solar system. Many objects were circling the Sun in that early era, and their orbits were such that they sometimes ran into one another. A Mars-sized proto-planet is hypothesized to have crashed into Earth about 4.5 bya. In the big collision, both were totaled; but gravity eventually pulled much of the flotsam and jetsam back together. Earth ended up being bigger, and another sizeable chunk formed into a satellite circling the Earth. It is a Humpty Dumpty story but with a happier ending. This, in fact, becomes a love story: Earth and its reconstituted satellite – the Moon, of course – became partners for eternity. (A recent hypothesis for the formation of the Earth-Moon duo does not rely on a collision but on a condensation of two different-sized bodies out of one cloud of cosmic flotsam and jetsam.) The gravitational connection that the two made had a very positive

The Gyroscope of Life

effect on Earth, because it stabilized the planet's tilt and rotational pattern as the two rocky bodies waltzed around the Sun. That in turn stabilized the planet's developing climate systems and made life on Earth a viable option.

Table 4.3. Key events in cosmology from The Big Bang to the appearance of life on Earth. Cosmic processes are thought to have begun near the time indicated, but some will continue into the future. For example, star formation may continue for another 100 billion years.

Billions of Years Ago	Event or Stage
13.8	The Big Bang
13.7	Stars begin to form
13.5	Milky Way begins to form
4.6	Solar system and Earth begin to form
4.5	Earth and Moon collide/co-condense
4.4	Liquid water develops on Earth
4.3-4.1	Life likely appears on Earth
4.0	Life leaves traces on Earth

Without a stable climate on Earth, it would have been very hard to get life going. Early living forms presumably hung on by their finger nails. It would have been especially hard for that life to get into a vital groove if the Earth were tilted at a much greater angle and the environment was more subject to drastic changes. Stability is a good thing, especially when trying something new. Think training wheels on a beginner's bicycle. Deep ocean vents, which are mineral-laden, hot-water geysers spewing out of the ocean floor, are sometimes mentioned as a possible breeding ground for life. They exhibit rather harsh conditions (high temperatures, high pressures, total darkness), but they produce a very stable environment. As a minimum, we know that lots of peculiar life forms now occur only within the vents' spheres of influence.

Populating the third rock from the Sun, and a biological proof of God

But we have gotten ahead of ourselves. Earth wasn't suitable for life 4.5 bya. The young planet was hot and volcanic with a toxic atmosphere. Although lots of water was present in the early atmosphere, it was all vapor. By about 100 million years after the big collision (or dual condensation) of Earth and Moon, things had cooled enough for liquid water to form. That set in motion the hydrologic cycle, which began to have Earth-altering consequences, forming oceans and rivers and sculpting landscapes by erosion. Sometime over the next half-billion years, life is thought to have appeared. The timing of that event is one of the most debated of the dates in Table 4.3.

Some have suggested that life was "seeded" on Earth by primitive, low-tech life forms arriving in meteorites. That is a plausible theory, but it begs the question of where and how life began. A favored theory for the genesis of life – wherever it happened – is that moderately complex chemicals, such as organic molecules in star dust, mixed with water to create a primordial soup. If it happened on Earth, energy radiating from the young Sun and lightning may have helped to form more complex chemicals that eventually developed a critical capability to self-assemble and produce more of themselves.

The Gyroscope of Life

It's a stretch – something that many biologists accept as an article of faith. No one has laid out a consensus-establishing sequence of events to get from naturally occurring molecules – no matter how complex – to an entity that can strike out on its own. No one has created conditions in which life has spontaneously developed. Still, many biologists assume that life self-organized here (or elsewhere in the cosmos) and then blossomed on what became an increasingly life-friendly Earth.

In the 1970s, I was convinced for a while that I had a biological proof for the existence of God. My proof was based on my understanding of the Central Dogma of Molecular Biology, which was first developed in 1958 by Francis Crick. A second, more popular version of the Central Dogma (CD) was published in 1965 by another of the co-discoverers of the shape of DNA, James Watson. The CD describes the flow of genetic information in living things. The shorthand version of Watson's statement of the CD was: DNA → RNA → Protein. If that seems a bit obscure, I'll elaborate. If it looks familiar, skip the next couple of paragraphs.

The blueprint for making an organism is encoded in its DNA. DNA is the abbreviation for the biochemical that holds all an organism's genetic details. Think of DNA as being somewhat like old-fashioned magnetic tape on which information could be stored as magnetic fluctuations. To retrieve the information, the tape was passed over a tape head, which could convert those magnetic variations into electrical impulses. Those electrical impulses were still encoded in a sense. Only with a decoding device, such as a speaker, could they be translated into a desired output, such as music. DNA's information is coded in chemical, not magnetic, form, but it still needs some sort of code-

Creeds, Creations, and Chronologies

transcribing system. That is the role of RNA. Parts of DNA's coded information – specifically portions constituting genes – are rewritten into a still-coded RNA form. Then the RNA's code can be translated or converted into a cell's workhorse proteins.

Table 4.4. Key events or stages of life as currently proposed. Life 1.0 is a presumptive original life form. Life 1.1 and Life 1.2 are Bacteria and Archaea. Life 2.0 (Eukarya) evolved and produced all higher life forms (plants, animals, and fungi). Many dates are subject to revision and are being revised regularly. For example, the "1.0?" stems from a very recent report of fossils of multicellular fungi found in 1-billion-year-old Canadian rocks.

Billions of Years Ago	Event or Stage
4.3?	Life 1.0 speculated to exist
4.0	Evidence of life in rocks and minerals
3.5-3.7	Life 1.1 and 1.2 evident in fossils
3.4-3.5	Photosynthesis appears
3.0-2.7	Oxygen-producing photosynthesis appears
2.4	Oxygen Crisis
1.8-2.0	Life 2.0 leaves clear evidence
1.2	Sexual processes evolve (for single cells)
1.0?	Multicellular life forms evolve
0.6	Stratospheric ozone layer forms
0.54	Cambrian Explosion of animal evolution
0.47	Plant and animal life develop on land
0.44	Mass extinction #1; 85% of higher life lost
0.36	Mass extinction #2; 75% of higher life lost
0.25	Mass extinction #3; 95% of higher life lost
0.20	Mass extinction #4; 50% of higher life lost

The Gyroscope of Life

0.13	Flowering plants appear
0.066	Mass extinction #5; 75% of higher life lost
0.002	*Homo* (the human genus) appears
0.00025	*Homo sapiens* emerges

In a somewhat weaker CD analogy, consider a non-French speaker looking through a French dictionary, copying down a definition into cursive form and then having a friend translate it into English. The protein (English definition) is being made or translated in a pattern dictated by a French dictionary (DNA), but cursive writing (RNA) was used as an intermediate. Going back to our magnetic tape example, magnetic codes from the tape (DNA) are selectively read by the tape head into electronic signals (RNA) and then transformed into music (protein). Again, in short: DNA ↠ RNA ↠ Protein: the Central Dogma.

But DNA ↠ RNA ↠ Protein isn't as simple as the arrows seem to indicate. It's not a strictly one-way process. It has loops. Proteins are needed to make DNA and RNA. It became a chicken and egg thing for me. All three of these complex biochemicals (DNA, RNA, and protein) had to be present "in the beginning" for the CD's system to operate. But how could such a complicated system have self-assembled from chaos? How could three very different components have arisen independently and then gotten together as the CD's trinity? It was like someone invented magnetic tape, while someone else invented a tape reader, and someone else invented a hi-fi amplifier with speakers. None of the three knew anything about the others' inventions. Yet, when their inventions were brought together, beautiful music happened. Very unlikely. Ockham's Razor suggested I invoke divine intervention – the original trinity, or maybe Roog or Pangu.

Creeds, Creations, and Chronologies

The CD has lost some of its scientific or theoretical clout in recent years. We now know that some viruses have no DNA. We now know that viruses and some living forms can make copies of DNA from RNA, i.e., RNA → DNA. We also now know that microRNAs, very short pieces of RNA and not genes, can control the protein-making process. In fact, microRNAs are causing the whole CD to be reexamined. Life is much more complicated that we thought. But, at the same time, these findings suggest that early life could have been much simpler. Many evolutionary biologists now think that the earliest life forms were RNA-based, with no DNA involvement at all. So, as the CD showed its feet of clay, so also did my biological proof for the existence of God. (While on the topic of clay, some have suggested that clay components in soil could have served as a template upon which complex chemicals assembled into self-sustaining and self-copying units.)

Hardware and software updates for life

Life left hard evidence for its presence on Earth by 3.5 bya. That evidence comes in the form of fossils found in Australia and Greenland. However, some clues suggest life may have been going on a half billion years earlier. Instead of working with fossil rocks, some archeologists study fossil chemicals. These are typically minute traces of odd substances in geologic layers – odd because they are complex chemicals thought to be produced only by living things. They are chemical fingerprints of ancient life. Chemical fossils in some very old rock layers provide good evidence that primitive life was present on Earth about 4 bya, and even more primitive life forms likely were around before that but failed

The Gyroscope of Life

to leave their calling cards.

Ockham's razor suggests life almost certainly was first practiced by organisms with a low level of complexity, with a simple technology – maybe with no DNA software. Maybe just RNA. We don't know what life 1.0 may have looked like 4.1 or 4.3 bya. Its practitioners must have been very small and maybe even non-cellular. But we know a major shakeup in life technology was coming. The chemical fossil fingerprints are faint and smeary, but they show two different kinds of technology, living hardware, were operating. We know what those two different kinds of life were because they are still around and still producing the same fingerprints today. One fingerprint was produced by Bacteria (or Eubacteria). The other belongs to the Archaea (or Archaebacteria). Archaea look much like Bacteria but use a somewhat different biochemistry or technology. Both are single cells surrounded by a membrane. Their genetic material is just a molecule of DNA sloshing around with everything else inside of the cell. But they had (and still have) distinctive hardware and software, and they left distinctive chemical fossil fingerprints.

Both Life 1.1 and 1.2 – Bacteria and Archaea – headed off down their separate evolutionary paths, and both became very successful. They constituted all living matter on Earth for about 2 billion years, and they still make up the great bulk of Earth's living material. Bacterial evolution produced versions equivalent to a Life 1.1.0.1, 1.1.1.4, 1.1.6, 1.1.8.3, and so on. Life 1.2 did the same, but the Archaea platform was perhaps more flexible. Species of Archaea are the dominant life form in some very harsh environments: thermal springs, deep ocean vents, and super salty water. The Archaea also developed some unique metabolic tricks – to include

Creeds, Creations, and Chronologies

making hydrogen and methane. Methane is the primary constituent in natural gas. It has accumulated in deposits over the eons because of the activity of Archaea.

Life 1.1 and 1.2 ruled the Earth for two billion years. "Ruled" is perhaps used with some irony. The earlier organisms were more likely just barely hanging on. But, as they evolved into more hardy forms and proliferated, they began to have impacts on Earth. Their metabolic processes produced mineral formations to include iron ore deposits. The Archaea produced methane, which is a potent greenhouse gas. That increased the greenhouse effect on early Earth, at a time when the Sun was 30% dimmer than it is today. The added methane – along with carbon dioxide from other early life forms – made Earth warmer and more life-friendly.

By 3.5 bya, some forms of Life 1.1 had developed an ability to capture light energy and use it to make complex compounds – the process of photosynthesis. Over the next billion years, some of those photosynthetic forms of Bacteria began releasing oxygen. This was a game changer. Until then, the atmosphere contained lots of carbon dioxide and very little oxygen. Adding significant amounts of oxygen to the atmosphere had many consequences. For one, the oxygen reacted with iron dissolved in the oceans, causing more iron-ore sediments to form. More importantly for then-living things (Bacteria and Archaea), the rising oxygen levels were toxic. Oxygen was a bane to life forms that had previously grown only in low-oxygen environments and had never evolved defenses against oxygen's ability to attack living matter. Many oxygen-sensitive bacterial and archaeon species disappeared around 2.4 bya when oxygen levels rose. This was the so-called Oxygen Crisis. On a positive note, the increased level of atmospheric oxygen was a boon to organ-

The Gyroscope of Life

isms that developed defenses against its attacks and that could use oxygen chemistry to produce energy. That chemistry included respiration, which is a key process in plants and animals today.

Some new kids were about to appear on the block occupied for so long by just Bacteria and Archaea. The first fossil-chemical hints of someone coming around the corner appeared maybe 3 bya, but these new kids didn't leave a distinct mark on the fossil record until about 2 bya. They seem to have evolved from Archaea, since they bear greater biochemical similarity to Life 1.2 than to Life 1.1. Life 2.0 (technically, the Eukarya or eukaryotes) produced larger and more sophisticated cells compared to Life 1.2. Those higher-tech cells had inner components that were enclosed by membranes. Key subcellular inclusions were: a nucleus to hold the genetic material, powerhouses called mitochondria that could take advantage of higher levels of oxygen, and chloroplasts that could carry on photosynthesis. Mitochondria are hypothesized to be specialized Bacteria that were shanghaied by Life 2.0 organisms. Chloroplasts are similarly thought to have been photosynthetic forms of Life 1.1 that were impressed into subcellular slavery by Life 2.0 forms.

Life 2.0 and its early descendants may not have been much to look at, but they were endowed with some advantages that were going to give them a big leg up on the evolutionary stage. That is not to say Life 1.1 and 1.2 became the underdogs. Bacteria and Archaea are far and away still the most diverse and successful forms of life on the planet – both in numbers of species and in the variety of places they can be found living. But Life 2.0, with its new kind of cell, was able to make living forms that a bacterium or archaeon could never dream of. After a couple billion years of evolutionary

Creeds, Creations, and Chronologies

experimentation, we now find eukaryotic forms as diverse as redfish and redwoods, peacocks and pea vines, cats and cacti, Venus Williams and Venus flytraps. Life 1.1 and 1.2 are still plodding along pretty much as single-celled, bacteria-like beings. They were and are very creative metabolically, but they don't have much of an eye for form.

The upstart Life 2.0 eukaryotic forms remained one-celled for over 1 billion years, but progress was being made one cell at a time in multiple areas, to include sex. Fossil evidence of a sexual revolution – seen in a type of cell division needed to make eggs and sperm – appears in rocks formed about 1.2 bya. Presumably, after that, living things could go to a single-cell bar and find some action.

Then another major milestone was reached around 800 million years ago (mya); multicellular forms of Eukarya appeared. Eukaryotic cells had finally learned how to cooperate with one another and act as a single unit – a huge step forward. With this advance, the stage was set for all today's higher life forms. It had taken 1.2 billion years of evolutionary trial and error (developing versions 2.0.1, 2.0.5, 2.0.9.2, etc.) to get cells with the 2.0 technology to play together – to develop cell-to-cell communication and coordination. That development proved to be a real breakthrough. In the 800 million years since the multicellular leap, life has moved from being just small groupings of cells to 3rd-, 4th-, 5th-, etc. generation plants, animals, and fungi.

By about 600 mya, photosynthetically generated oxygen began to form a stratospheric ozone layer around Earth. The upper-level ozone band reduced ultraviolet (UV) radiation arriving at the surface. UV presents little problem to aquatic organisms, since water absorbs the energy of UV; but living on land without some special UV shield would have played

The Gyroscope of Life

havoc with land pioneers. DNA is quite susceptible to damage from UV, which is abundant in unfiltered sunlight. The ozone layer provided a UV filter and made it much safer for terrestrial forms to eventually evolve onto land.

Somewhere between 600 and 500 mya, Earth experienced an evolutionary supernova. The Cambrian Explosion marked a time when aquatic animals rapidly multiplied and diversified. A lot of the change was achieved by small evolutionary steps made over hundreds of generations. But some of the evolutionary products from the Cambrian Explosion seemed almost to be made of whole cloth. Most of today's major animal divisions appeared.

Primitive plants moved onto land about 430 mya. Initially, they were little more than green algae that could live in damp places. Fungi moved onto dry land about the same time, and that may not be coincidental; the two different life forms may have struck a deal that helped them mutually achieve terrestrialism. Before long, green plants covered the Earth. Coal formation and natural gas deposits followed.

A last major milestone for plants was the appearance of flowers. The first indication of a flower in the fossil record is from about 130 mya. A period of rapid expansion of flowering plants occurred about 40 million years later. Familiar groups such as the ones containing roses and magnolias evolved their distinctive characteristics during this era. The grasses underwent a similar burst of evolution about 35 mya, and grasslands became a major global vegetation form by 10 mya.

Mixed in with the evolutionary booms came some major biological busts. Paleontologists – hybrid geologists-biologists – find in the fossil record evidence of five episodes of "mass extinction" (Table 4.4). These are cases where fossils of

Creeds, Creations, and Chronologies

many plants and animals are suddenly no longer present in newer, higher geological layers. The largest of the five mass extinctions occurred about 250 mya and was suspiciously coincident with the eruption of a super-volcano. In short order (as reckoned in a geological or evolutionary time scale), 95% of the then-living plants and animals dropped out of the fossil record. They are there in older layers, but they are suddenly missing in layers just above. That mass extinction of 250 mya was almost like a total reboot for life. That must have dictated a decided plot turn in life's improvised play.

Proof? Sorry, science only offers evidence

Tables 4.2 and 4.3 provide an impressive array of cosmological dates and events. What is the evidence for any of it? Why do cosmologists theorize the universe was born in an instant 13.8 bya? How do they determine whether a star will burn down or blow up? They use known physical principles, computer models, data from particle colliders, astronomical observations, and Ockham's razor (among other things) to develop theories. Their theories mathematically help explain how the universe could evolve from an infinitely small and infinitely energetic point – even if they cannot yet explain how such a point would come to be. Such theories are not testable. But, then, science is not in the business of proving good theories or hypotheses – only disproving faulty ones. Cosmologists offer a lot of evidence to suggest their theories are on the right track.

What are the proofs for the dates and events in Table 4.4 that pertain to life and living things? There are none. Biologists cannot prove a spontaneous origin of life nor can they prove a 3- or 4-billion-year progression of life, but they can

The Gyroscope of Life

offer a slew of evidence to support at least the second of those two propositions. Well, it's not just biologists. The other half of natural history is very much involved in developing the evidence and especially in putting dates on events. Think of those fossils in the National Museum of Natural History. Geologists are at least equal partners in putting together the pieces that become those reconstructions, and we rely solely on them for determining how old the fossils might be.

The dating of fossils is really a pretty straightforward process – once all the groundwork has been laid. Geologists who first studied layers of rocks and sediments in the Earth's crust theorized that the deeper layers were laid down first. (Well, maybe that was more an article of faith than a theory to be tested initially.) They also observed that the plant and animal fossils in older/lower rock layers were different than those in newer/higher formations. It seemed suggestive, but they really had no way to know just how old any of those rock layers were.

The 20th century saw the appearance of techniques for determining the age of geological materials. Radioactive or radiometric dating methods now allow geologists to establish – within a range – the age of rocks, fossils, artifacts, etc. Depending on a number of factors, the range estimated by a radiometric measurement can be accurate within 1% or less. For example, using uranium decay to date a rock sample can establish an age of up to 2.5 billion years within a plus or minus of only 2 million years – less than 0.1% error. Such techniques put a pretty definitive age on rocks and the fossils found in them. With radiometric dating – definitely not to be confused with speed dating – we can create a timeline to go with fossil evidence. The fossil evidence is interpreted by hybrid geologists-biologists: paleontologists.

Creeds, Creations, and Chronologies

The fossil evidence is very sketchy for 1 or 2 billion years after life first appeared, but chemical evidence in 4-billion-year-old rocks suggests that something was making complex biochemicals back then. Rocks that are 3.5 billion years old show signs of photosynthesis going on. Fungi are clearly evident by 1.5 bya. And so on. I will not elaborate every milestone event or stage, nor will I belabor the nature of the fossil evidence for each stage or event. Suffice it to say that the former natural historians – geologists and biologists – are in general agreement about the evidence supporting the big picture.

Many of the events and timelines proposed by science for the formation of the universe, Earth, and life are based on observations, calculations, or theories that can be reexamined, recalculated, and revised. Scientists tend to feel comfortable with theories that align well with current understandings about universal laws and principles. The universe, stars, and planets are presumably inevitable consequences – simply aftermaths – of The Big Bang and the physical laws that govern matter and energy. Life's appearance on Earth was either serendipitous or supernatural, but life's eventual occurrence was perhaps inevitable somewhere in the universe because of presumably universal physical and chemical principles. Ten to 80 sextillion (that's 21 zeros) stars are out there in the universe, and many of them likely have satellites that could provide favorable environments for self-propagating forms. With such numbers, the odds of life developing in multiple forms and places would seem to approach certainty.

Biologists have great confidence that natural selection and evolution are driving forces for proliferation and diversification of life on Earth. They can point to evidence that evolution is happening in real time – not just citing fossil evidence

The Gyroscope of Life

that suggests it must have been working in the past. Evolution and natural selection are such important themes that a later chapter is devoted almost entirely to them.

5

House Rules and Theatrics

All the world's a stage…

William Shakespeare (1564-1616)

To be or not to be…

If the fossil record is any indication, many once-living things have failed at life. Think dinosaurs and sabretooth tigers. But, as we look around today, some living things appear to be doing well. Think cockroaches and crabgrass. Can we study biological failures and successes and figure out a formula for success – Mother Nature's House Rules for survival? I think we can. I am going to give it an autobiological shot anyway, but let's set the stage first.

We can safely assume that 99.99+% of the individual plants, animals, fungi, etc. that have ever lived are now dead, and everything currently alive will eventually shuffle off this mortal coil also. If biological success were to be defined as individual immortality, we would be off on a fool's errand in search of successes. In fact, immortality is apparently undesirable from Mother Nature's standpoint. She seems to like new players stepping onto the stage where life's improvised dramas play out, and it could be problematic if has-been players failed to exit. So, she has built in a hook: mortality.

From the standpoint of those of us now upon the stage, mortality has been a good thing. It is good that 99.99+% of once-living things are dead. It's not so much that we would

The Gyroscope of Life

have to worry about being eaten by *T. rex* or sabretooth tigers. Rather, it's about the space and resources needed to live. Without mortality, the room and raw materials needed by new organisms would have eventually become tied up by undying things. When organisms die, the minerals that make them up can be recycled into humans, horntoads, and other new players. Without death and recycling, we can suppose evolution would have ground to a halt well before humans stepped on the stage. Mother Nature would have only the same monotonous play to watch endlessly, and the stage would be jammed with players who refuse to exit. But she has a hook…

So, if biological success and failure cannot be gauged within mortal individuals, we must look to the groupings that we call species. More will be said about the notion of species in another chapter, but begin to think of it as a concept to which Mother Nature pays attention. She appears interested in the development of new players – what we would call new species – in life's unscripted play. She does not always color entirely within species lines, but she does grant a measure of longevity, if not immortality, to the groupings we call species. They are the actors for whom she wrote the soon-to-be-enumerated House Rules, and they are the entities that we can consider in our search for biological successes and failures. In a sporting metaphor, a species is one of the myriad teams in life's chaotic game. Each team has a variable number of players, but when no more players are around to check in, "extinct" is posted after the team's/species' name.

It is likewise safe to assume that 99+% of the species that have ever lived are now extinct. They have left the stage (or the playing field). The fossil record tells us that, for higher-

House Rules and Theatrics

life forms (plants, animals, and fungi), extinction is a more likely outcome than is survival. The Cambrian Explosion, which began about 540 mya (Table 4.4), saw the rapid appearance of several major animal groups, but some of them quickly disappeared. They became evolutionary dead ends – actors that just didn't suit Mother Nature's tastes. Most of them did not survive the mass extinction that occurred 440 mya (Table 4.4). If they did, they still faced the mass extinction of 250 mya in which 95% of the then-alive plants and animals disappeared. Extinction is more the rule than the exception.

The modern-day nautilus seems an interesting exception to the "extinction rule". Fossil nautili from 500 mya look like the nautili swimming around today. Accordingly, nautili are sometimes described as living fossils. However, none of those fossils is considered the same species as any of today's nautili – not even in the same genus. Same story for 450-million-year-old horseshoe crabs and 400-million-year-old coelacanths (a type of fish). Today's forms of these animals look much like their fossilized ancestors, but none is considered to be in the same genus – let alone species – as the ancient forms. Evolution seems to be running in slow motion in these groups, but it is moving ahead. The ginkgo tree is another living fossil. Fossils of trees that look like today's ginkgo date back 270 million years. But, to be clear, all plant and animal species that were alive 200 mya or more are no longer with us. With less certainty, we might suppose the same is true for species that were alive 10 mya. New and improved versions or reconfigurations may be around, but those ancient species are no longer prancing about the stage. The theatrical lifetime of actor species is not infinite.

The timings and causes of some species' extinctions are

The Gyroscope of Life

generally known. Most of the last dinosaurs disappeared 65 mya in the last mass extinction, which was likely caused by an errant asteroid. The last reliable sighting of the dodo was in 1662, just 24 years after first human settlement on the Mauritius Island home of the big, flightless birds. We know when and why (primarily over-hunting and habitat loss) passenger pigeons went extinct. The last-known member of that species, which once numbered at 3 to 5 billion, died in 1914. The American chestnut tree has been brought to the brink of extinction because of the accidental introduction of a disease-causing fungus from Asia. For these and many more now-extinct species, the individuals and their gene pools have disappeared. The hook has pulled those actors off the stage. No more players are left to check into the game.

Here is the $64 million question. Can we humans learn from the extinct 99+% how to avoid their fate? Or is "99+%" the handwriting on the wall for all of us? Is extinction really a rule? We can perhaps take some heart in the search for extinction-avoiding strategies when we note that some of the 99+% weren't annihilated by environmental catastrophe or a doomsday disease. Rather, their traits survived as their offspring's offspring gradually evolved into new species. Such superseded species did everything right as far as the soon-to-be-introduced House Rules were concerned. But over time, the collection of mutant traits in their gene pools made them look and behave differently. Mother Nature is generally okay with such genetic makeovers. In fact, life seems to thrive on them. That's what happened to nautili, horseshoe crabs, and gingko trees. Through evolutionary time, a successful species is no longer the same actor. Accumulated changes have made it physically and genetically different from forbears 500 or 10,000 generations removed;

House Rules and Theatrics

it has evolved into a new species. These are notions that were mentioned as part of a biologist's credo. We can reason that those forbears were biologically successful even though their species graded into extinction. We human beings are hypothesized to have descended (ascended, as some would have it) from extinct hominid species whose genes we still carry.

So, biological success might be defined as the long-term survival of a species and its gene pool. But, given our own species' short history (maybe 250,000 years), "long-term" seems a bit of a stretch for describing our time on the stage. From Mother Nature's 4-billion-year perspective, even 100 million years is not a long time. Consequently, for our purposes, biological success will be gauged simply by whether a species is one of today's actors. If it is, it has presumably been doing key things right – obeying the House Rules.

After toying with this notion for many years, I am ready to put words in Mother Nature's mouth and offer autobiologically what I think might be her House Rules for biological success – for survival. There are only three:

1. Fit your environment; adapt. (An ecological imperative)
2. Keep your entropy low; get and keep your shit together. (An organizational imperative)
3. Pass on your DNA; make love and more of yourselves. (A reproductive imperative)

We will take up each house rule in turn, describe how it applies, and illustrate how it is crucial to the long-term, extinction-avoiding success of a species.

About Mother Nature (and alternative rule makers)

The Gyroscope of Life

The notion of a female deity who oversees the Earth and its creatures is found in the traditions of many peoples. In ancient Greece, she was Gaia, Mother Earth. In Roman times, she was Terra. In other cultures, similar goddesses played a maternal role to living things and their environment. While I am at best agnostic about an Earth goddess, invoking such a being has a literary appeal when discussing biological sequences and outcomes – causes and effects. Hereafter, I will call her Mona, a nickname by conjunction of Mother Nature. Mona is not a stereotypical mother – not a particularly protective nurturer. From all indications, she is more like a tough, crusty, and uncompromising stage manager. No mollycoddling. No excuses. No second chances. She plays no favorites. To all of her children, she says, "These are the House Rules. Screw up, and you are history."

If you do not believe in Mother Earth or Mother Nature but hold that a vitalist deity is inspiring life, I would suppose you might still believe there is some rhyme and reason to the appearance and disappearance of species. If so, please feel free, as the spirit moves you, to think of God, Roog, Pangu, or another life-granting deity as the author of the House Rules. Or, if you are a pure mechanist/reductionist, you might simply insert "Life's" in front of "House Rule".

Mona's House Rule #1: Fit your environment; adapt. (An ecological imperative)

We must begin by making it clear that Mona is using "fit" as a verb in HR#1. Why the grammar lesson? Because fit, fitness, and fittest are often used by biologists as nouns or adjectives to describe a different concept – "survival of the fittest". We will be using fit in that nominative sense later.

House Rules and Theatrics

But in HR#1, fit is an active verb; fit your environment is a verbal command. Organisms must fit, or be adapted to, their environment. We'll get into some specifics of fitting one's environment, but let's first establish what "one's environment" is.

We often use "environment" loosely to describe the stuff around us, with the boundaries of "around" being equally nebulous. Mona uses environment in a more specific way in HR#1. The word conveys functionality and boundaries: all the features and properties in an organism's vicinity that affect – either positively or negatively – its ability to live there. The space that constitutes an organism's environment can be extensive, especially for organisms that are mobile. In a final chapter, we'll introduce the concept of ecosystem. HR#1 could equally be rendered as "Fit your ecosystem."

By Mona's definition, features and properties in an organism's milieu that do not affect its ability to live there are not part of its environment. Hair-splitting perhaps, but we need a clear understanding of what environment means, since understanding HR#1 is a matter of life or death. Organisms can ignore – do not have to be fit for – features in their space that do not affect their livelihood. These are things like some biologically inert gases in the atmosphere, minerals in the water or soil that do not affect life, and which direction the wind blows. These are not part of environment as used in HR#1. They require no action by organisms struggling otherwise to obey the House Rules. Conversely, some environmental factors may not be present continuously in an environment, but they must be taken into account by organisms trying to fit in their ecosystem. Factors like diseases, fires, floods, and droughts occur periodically (sometimes yearly, sometimes once a century) in many environments, and or-

The Gyroscope of Life

ganisms must be adapted to, or fit for, such challenges.

I will use environment in Mona's organism-specific way, but I will also use it on occasion to distinguish between different environments. If Mona were to use it in that distinguishing way, she might say something like, "I don't care which environment you choose, but you'd damned well better fit it." Those of us with no maternal or managerial responsibilities for life might say something like, "this environment provides a better fit for such-and-such organism than that one does." But, if Mona heard us say that, she would be in our face. She would never describe the relationship between a living thing and its environment in that way. The environment is not an actor. It is the stage. It does not provide a fit. Rather, the onus is entirely on living things to accommodate, or fit, themselves to their environment. If they are going to be successful, they must actively adapt in such a way that their basic needs can be met by the environment as it is. Fitting in, or adapting, is an active process carried out by living things.

If Mona wrote fairy tales for her children, the Goldilocks story would go something like this: "Goldilocks goes into a bear's house where she finds the porridge is too cold, the chairs are too big, and the beds are too hard. And then the bear eats her. The end." Moral of the story: the environment is what it is; live within it (or die). A living thing has to be adapted (or adapt itself) to its environment. It must make-do on cold porridge, big seats, and hard beds if that is all its environment offers.

Earth offers a kaleidoscope of environments for life, and, wherever we find life, we can know that Mona's imperative about fitting is being met. Seen through the eyes of a biologist, the amazing diversity of life found in diverse environ-

House Rules and Theatrics

ments is a testament to evolution and natural selection – to Mona's wisdom in being so hardnosed about HR#1.

Not surprisingly, plants and animals that are well-adapted to tropical rain forests tend to be abject failures in deserts, and vice versa. Species that can withstand the drought, blazing days, and cold nights that characterize deserts cannot handle the wet, gloomy mugginess inside a rainforest. Similarly, organisms that live happily in mountain springs cannot survive in equally cool but salty tidal pools. In accommodating themselves to fit into one type of environment, they forego life in others. Adaptation – fitting into a particular environment – is a two-edged sword.

And here comes the sucker punch. An environment that an organism has accommodated itself to and that has been suitable for millennia can turn murderous in the blink of an eye. Things that plants or animals have never experienced evolutionarily – like asteroid impacts, new diseases or predators, rapid climate change, mall constructions, clear-cuts, pesticides, etc. – can wipe out populations of organisms that have lived on their little pieces of the Earth for perhaps millions of generations. What is Mona's response? See HR#1.

The story of the demise of the American chestnut illustrates both success and failure under HR#1. The American chestnut has largely disappeared because of a fungal disease introduced from Asia. Japanese and Chinese chestnut trees (closely related to American chestnut) are susceptible to the fungus, but over the centuries, they evolved some resistance to the disease. They developed genetic solutions, mutations that granted them a large degree of immunity. So, they were able to fit and survive in their fungus-infested environment.

But, when Asian chestnuts were imported into the US in the late 19th and early 20th century, so was the fungus.

The Gyroscope of Life

American chestnuts had no resistance to the disease; they had no evolutionary history or experience with the disease – and no mutations that might grant them some immunity. HR#1 suddenly loomed large for American chestnut. These were trees that represented maybe 25% of the hardwoods in the Eastern US. They were a dominant species in all respects, with trees up to 100 feet tall and trunks that could be 14 feet in diameter. They were highly prized for their lumber, because a typical tree could produce five, 10-foot, rot-resistant, knot-free logs. I've seen a few young chestnut trees still hanging on here and there, but they lack all of their former glory. They succumb to the disease before they get large. All this because the species was suddenly no longer able to fit its (altered) environment.

Foresters and plant pathologists are trying to help the American chestnut regain its footing. They are breeding the Asian chestnuts' disease-resistance genes into American chestnut. If their efforts are successful, it's possible that American chestnut could again be wonderfully well-adapted to the forests of Appalachia, where it was the signature species and something of a cornucopia – providing rot-resistant wood for homes, barns, fences, railroad ties, furniture, cabinetry, and caskets. And the fruit – the chestnuts – were a cash crop, food, and animal feed for many Appalachian and Southern families.

A major, potentially fatal alteration to one's environment is not always the kiss of death, though. Through fortuitous mutations and natural selection, some organisms are able to "re-fit" themselves – as with the Asian chestnut species. Some dinosaurs were able to adapt to and evolve in environments drastically altered by an asteroid's impact; modern birds evolved from feathered dinosaurs. But Mona would seem to

House Rules and Theatrics

be only in the earliest stages of working with her children on adapting to bulldozers, chainsaws, and rapid climate change. Organisms faced by those modern juggernauts are at great risk of no longer fitting their environment, because it can change so much faster than they can.

We've determined what Mona means by environment and fitting it, and we know what her penalty is for not fitting. But to whom exactly does HR#1 apply? The answer, in short, is all living things. Any organism that finds itself out of synch with its environment is thereby at great risk of dying. But every individual organism is going to die anyway; Mona has granted no one immortality. For any individual organism, the penalty for violating HR#1 is hastened mortality, and, from the big picture standpoint, that is at best an "uh oh".

For an entire species, though, wholesale violation of HR#1 is a definite "oh, shit!" If a species' environment becomes so altered that none of the team can fit in, the team will end up with "extinct" after its name. It will become an evolutionary dead end. Its gene pool will be drained. None of its genes will pass to a new and improved species. The hook comes out, and the actor disappears. All because it was unable to obey HR#1, fit your environment. HR#1 is an ecological imperative. We will have more to say about ecology in our last chapter.

Mona's House Rule #2: Keep your entropy low; get and keep your shit together. (An organizational imperative)

Mona's second House Rule is a bit more difficult to understand and just as tough to comply with. To appreciate

The Gyroscope of Life

how tough, we need to delve briefly into thermodynamics. Thermodynamics is a branch of science and engineering that was developed to deal theoretically and mathematically with how heat and temperature are related to mechanical work and energy. It started out as an esoteric field, providing information that was actionable for engineers working with mechanical systems – steam engines and the like. However, chemists discovered that thermodynamic concepts could also be applied to chemical systems. That put thermodynamics into the sphere of biology – since living matter can be viewed as a collection of chemicals and chemical processes. Understanding the thermodynamic rules of the road for chemical processes can help us better understand life. And, as a bonus, some thermodynamic terms can be fun when dropped into cocktail party conversations.

Thermodynamics is encompassed in four Laws. These are not go-directly-to-jail-do-not-pass-go laws, nor are they like Mona's drop-dead House Rules. Rather, they are universal principles, or concepts, much like the Law of Gravity. Nothing can "violate" the Law of Gravity. We know it always works even if we don't know how. The same is true for the Laws of Thermodynamics. They are statements or observations about perpetual truths or phenomena. They simply describe how things are, albeit in some pretty abstruse terms.

Two thermodynamic Laws are of particular interest in understanding the bind into which HR#2 puts all living things. The 1st Law is a good one actually, one we can all live with. In fact, it is one we cannot live without. It can be stated in various technical ways, but a simplified version says "energy can neither be created nor destroyed, only changed in form". This is sometimes called the Conservation-of-Energy Law. It could also be called the Interchangeability-of-Energy Law.

House Rules and Theatrics

The 1st Law suggests that the amount of energy in the universe is fixed. Energy is not being made – not since The Big Bang. It is continually being used, but it doesn't get used up in the process; it is only changing from one form to another. Think of it as a recycling system where the amount of energy is fixed, but the form the energy takes keeps shifting as it cycles and recycles through the system. That's a lot to wrap one's head around, especially since energy is a difficult concept all on its own. Energy provides the effort or force needed to do work, which is another fraught term. It may help just to look at some examples of the things that happen when energy is being used or transformed and work is being done.

In mechanical systems, energy can be packaged in several forms or categories. Kinetic energy describes the energy possessed by an object in motion. When a hammer strikes a nail, the hammer's kinetic energy is transferred to the nail, driving it in, and doing work. Heat is another form of kinetic energy that can be used to do work. Heat causes molecules to vibrate more rapidly – more energetically. Steam engines use heat's kinetic energy (in the form of water vapor molecules that vibrate violently) to do the work of moving pistons and shafts. To understand where the energy needed to heat up water into steam comes from, we need to consider another major category of energy forms – potential energy.

A rock at the top of a cliff represents potential energy. If it is pushed off the cliff, its potential energy is rapidly transformed into kinetic energy. In our steam engine example, potential energy was used to generate the heat that boiled water and made the steam. More specifically, it was energy that had been stored in the chemical bonds of wood or coal – chemical energy that was transformed into heat or kinetic

The Gyroscope of Life

energy to drive a piston. Because biological matter is chemical matter, biologists' ears perk up when chemical potential energy is the topic. It deserves closer consideration.

Chemical energy is the energy stored in the bonds between two molecules or atoms. Think of it as a force or power that must be continually exerted to hold the two together – like a rubber band stretched around the two to bind them together. If the bond is broken, as can happen when wood or coal is burned, the bond's energy is released – often released as heat and light. This is totally okay according to the 1st Law. Energy is neither being created nor destroyed, only changed in form – from chemical energy to light and heat, both of which are just other forms of energy.

But where does the energy in chemical bonds come from? We can be sure of one thing; it was not created to form the bond. Energy isn't being made anymore. Rather, energy must be transferred and transformed from other sources to establish the chemical bond. In a test tube, the energy causing chemical bond formation is the kinetic energy of heat. Holding the tube over a burner makes the molecules inside it vibrate more rapidly. When two warmer, faster-moving molecules bump into one another, enough energy may be provided in the collision to cause them to combine chemically—to form a chemical bond. The heat or kinetic energy that had been jostling the molecules is captured in energy-rich chemical bonds. It's the 1st Law at work. Kinetic energy was transformed into chemical energy, and work was done.

One of the most important thermodynamic events in biology is a happy consequence of the 1st Law. In photosynthesis, the radiant energy in light is transformed into chemical energy. Energy streaming in from the Sun is captured by plants and used to form energy-rich bonds in a

House Rules and Theatrics

variety of chemicals – chemicals and energy that can then be passed through food chains to all living things. Then, when the chemical bonds are broken in the innards of whatever organism the chemicals might have ended up in, the energy released can be used for many purposes. Such purposes can include forming more chemical bonds (to make DNA, for example), making light (if the organism is a lightning bug), or generating heat (if the organism is warm-blooded). Each transformation of energy in this sequence – from sunlight, to chemical bonds, to heat or light – is simply obeying the 1st Law. Living things use the 1st Law with abandon. It allows us to obtain energy in various forms from our environment and then convert it into forms that are more immediately useful. In short, we benefit greatly from the 1st Law. Life couldn't happen without it. It allows living things to get their shit together.

The 2nd Law of thermodynamics is more of a problem for living things – almost a kiss of death. It can be stated in difficult-to-absorb ways, but one of the most straightforward wordings is deceptively simple: "heat cannot spontaneously flow from a cold location to a warm location". That's like saying water cannot spontaneously flow uphill. Everybody knows that! And that is exactly what makes this a Law. It is a first principle, merely a statement of how things are. If we dig a little deeper, the 2nd Law becomes less obvious and more ominous, even sinister. When taken to a logical conclusion, the 2nd Law suggests that gradients in energy levels between two areas will eventually and inevitably even out, with the higher levels of energy flowing into the lower. Heat cannot spontaneously flow from a cooler location to a warmer location, but it will inevitably flow from warm to cold. So also for chemical energy. Chemical energy will not

The Gyroscope of Life

spontaneously flow into and build up energy-rich chemicals. Rather energy-rich chemicals will inevitably break down, release their energy, and become simpler. With time, the 2nd Law drives energy and matter toward evenness or a uniformity. Even the best insulated thermos bottle is eventually going to end up with a drink inside that has cooled down to or warmed up to the ambient temperature.

The impact of the 2nd Law is something we know intuitively or by observation. In a trivial, non-thermodynamic example, we know that, no matter how neat we make a room, it will get messy. After we put energy into creating neatness and order, just living there inevitably creates disorder. The only way to keep the room neat is to keep putting effort/energy into it.

But let's get biological and truly thermodynamic with our discussion of the 2nd Law. When things die, they quickly began to look different. The complex matter of which living things are made begins to return to simpler, less complex forms. That's because of the 2nd Law. Bacteria, fungi, and worms can speed the process along, but, even in a germ- and worm-free environment, energy-rich matter will gradually fall apart into simpler, lower-energy stuff. It's the dust-to-dust thing. All living matter is assembled – using the life-friendly 1st Law – into complex forms from bits and pieces of exploded stars, but once an organism dies, all that complexity will return to a star-dust state. Likewise, the chemical energy that has been holding all of the complex matter together will be released back into the universe.

Thermodynamics has a term to describe and quantify this inevitable movement towards oneness with the universe: entropy. Mona uses the term prominently in HR#2 – Keep your entropy low. Entropy is a measure of the disorder or

House Rules and Theatrics

randomness of energy and matter. Disorder is what living things must keep low. And it is a great word to drop into a conversation at a cocktail party, because it has evolved not-strictly-thermodynamic connotations. "My desk has been overwhelmed by entropy." "This field of candidates is generating lots of entropy." "Entropic forces seem to have taken hold of the team during halftime."

Entropy is more about what isn't than what is. It is a measure of disorder or dispersion. Entropy's numerical value goes up as order and energy levels go down. It's like measuring the distance between cars to determine how uncongested the road is – rather than just counting the cars. Despite being conceptually a little tricky, entropy is a useful, if ominous, concept. When we introduce entropy into the definition, the 2nd Law of Thermodynamics says that, when left to their own devices, isolated systems will always move towards higher entropy, or disorder. It's the thermodynamic equivalent of Murphy's Law – anything that can go wrong will go wrong. It is technically related to the notion that shit happens. The 2nd Law makes it difficult for a living thing to obey HR#2's requirement to keep its shit together.

Think of a living thing as a spinning top that is made of loosely attached components. The organism does well as long as the top spins smoothly. But as this metaphorical vital assemblage spins, metaphorical parts can fly off. Loss of those parts can throw the top out of balance, destabilizing it, and causing it to wobble or even teeter over. That's what the 2nd Law says will eventually happen to living things. That's what entropy does. It flicks pieces off the top. They must continually be replaced to keep the top – life – spinning smoothly. Fortunately, the 1st Law allows for that continual repair, and even for growth. Organisms obtain energy and matter from

The Gyroscope of Life

their environments and use those inputs to keep the metaphorical top fully assembled and balanced. Energy inputs keep it spinning and also provide for repair of parts that might fly off. In so doing, a living thing meets the requirement of HR#2; it keep its entropy low. It is quite literally a do-or-die achievement.

But, remember the House Rules are really addressed to species – not individuals. There is no great biological harm if a few members of a species fall into entropy's maw. As long as enough members of the species keep their stuff together and as long as they are mindful of the other House Rules, the species will stay in the game. The actor will remain on the stage. But, if things deteriorate to the point that no members of the species can stay away from entropy long enough to reproduce, HR#2 has been violated. The penalty is extinction. HR#2 and the 2nd Law can be a murderous duo.

We can think of HR#2 as an organizational imperative. To remain alive, things must establish and maintain a critical order, which the 2nd Law is continually working to destroy. We will have more to say about life's organizational issues in the next chapter.

Mona's House Rule #3: Pass on your DNA; make love and more of yourselves. (A reproductive imperative)

This is the easiest of Mona's House Rules to understand and obey. In fact, individuals of many species clearly take pleasure in obeying it. But, as with all three House Rules, it is aimed primarily at entire species, not individuals. It is unique among the three House Rules in that an individual

House Rules and Theatrics

organism can violate it without adverse consequences to its own life span. An organism that fails to pass on its DNA gives up the potential for some measure of genetic immortality, but it incurs no other downside. In fact, some organisms can prolong their lives by avoiding reproduction. For some plants (century plants and many of our crops) and some animals (salmon and some insects), reproduction is life's climax and leads inevitably to death of the procreants.

HR#3 may be the rule that Mona monitors the most closely. It might even be her equivalent of a Prime Directive. If DNA is being passed on, the other two rules must have been followed also. Nothing else has to be passed along from generation to generation. The rest of the plant or animal can return to star dust when the individual dies. The DNA molecules in an egg and a sperm provide the blueprint – the business plan – to make a copy of the parents.

The case can be made – and was made eloquently in 1976 by Richard Dawkins in his *The Selfish Gene* – that all organisms in all their diversity and complexity are merely replicating engines for DNA. The chicken is just Mona's way of making another egg. There are no rewards or bonus points for being beautiful or different – only for being good at making copies of one's DNA and passing them along. Species that screw that up – that waste effort on things that won't increase their odds of passing along their DNA – are not going to make the cut. But, if their babies turn out to be better carriers of DNA, bonus points go to the species. Mona knows that the very act of passing on DNA carries with it a chance for genetically improved life forms and eventually new life forms, and new players and new acts seem to please her. Perhaps that is why she makes the DNA-passing process so pleasurable for some of her children. We

The Gyroscope of Life

describe HR#3 as a reproductive imperative. We will have more to say about reproductive strategies and kinkiness in another chapter.

6

Dead or Alive?
And the Gyroscope of Life

The reports of my death have been greatly exaggerated.

attributed to Mark Twain (1835-1910)

Binary thinking

As a student, I liked true-false tests. They were easier than multiple-choice tests. A multiple-choice question with five choices was like a set of five true-false questions, all of which I had to get right to get credit for answering just one damned question. A true-false question has a reassuring binary quality about it; it's either right or wrong. Binary properties perhaps appeal to us at an intellectual level. There is a tidiness and absoluteness about knowing what we are considering has only two mutually exclusive possibilities. It's either this or that – not an indefinite number of options. Some tidy binary-seeming qualities include: 0 or 1 (in base-2 math and digital electronics), no or yes, down or up, male or female, and dead or alive.

But Mona does not seem to have a strong preference for binarity. A few qualities in nature are binary as far as we can tell from our four-dimensional perspective. Examples are right or left, having mass or not, positive or negative (if electrically charged), and at rest or in motion. But some qualities that might seem binary become more complex on closer

The Gyroscope of Life

examination. One of those is dead or alive.

Being dead or alive would certainly seem to be a binary quality. Common sense suggests that everything should either be alive or dead – with the great, great majority of matter falling into the latter category. But one doesn't get far into the biological world before discovering a fascinating fuzzy area between dead and alive. We'll consider first some nonliving things that have attributes we associate with life. Then we'll consider some life forms that can play dead quite convincingly.

Prions: Scary, weird protein hackers

Prions (pronounced pree'-ons) are simple, nonliving entities that act as though they are alive. They are disease agents, but nothing like the usual suspects: bacteria, fungi, viruses, and parasites. Prions infect animals, cause horrific diseases, make copies of themselves, and then disperse to start more infections. Prions cause mad-cow and mad-cow-like diseases in domestic and wild animals and all-too-similar diseases in humans. All known prion-caused diseases are fatal. All attack neurons (nerve cells) of the central nervous system. First symptoms are relatively mild behavioral changes, but the victims move rapidly to disorientation, loss of motor control, dementia, and death. The definitive diagnosis for prion-caused diseases can be made only post mortem; the brains of victims are riddled with holes and quite spongy looking.

Kuru is a horrific, prion-caused, human disease. Mercifully, it has been eradicated. It produced all the prion symptoms noted above, to include brains that look like sponges. It occurred only in Papua New Guinea within an indigenous group called the Fore People. The first case of kuru was re-

Dead or Alive?

ported there about 1900. By the 1950s and 1960s, as many as 300 of the 20,000 Fore People were dying of kuru each year, and the toll was highest among women and children.

In the early 1960s, well before the causal prion was discovered, the path of kuru transmission was identified. With that information, the disease cycle was easily broken. The Fore People practiced ritualistic funerary cannibalism. The name pretty much says it all. It was a Fore custom to cook and eat deceased relatives to honor them and to free their spirits, putting an entirely different spin on family barbeques. Children and women were chosen to eat the honoree's brain, which turned out to be the richest source of kuru-causing prions. With the mode of transmission established, prevention of kuru was literally a no brainer. Funerary cannibalism disappeared in the 1960s, and the last death from kuru was reported in 2005.

We now know that prions are simple protein molecules. Discovering that was quite a feat scientifically. Stanley Prusiner earned a Nobel Prize for his work in the 1990s, proving that the infectious agents for these brain-eating diseases, to include kuru, are just a couple hundred amino acids strung together. In other words, they are small proteins. Prusiner coined the word prion from a conflation of protein and infectious. As the story unfolded, prions sounded like something from science fiction in which a submicroscopic, brain-eating robot becomes a self-replicating scourge. But these non-living, disease-causing agents are real – dead but deadly. Prions cause disease in much the same pattern as viruses and bacteria: infecting, making copies of themselves, dispersing, and starting over in another host. However, there is a major distinction. Viruses and bacteria carry genetic instructions (DNA or RNA) for bringing about infection and

The Gyroscope of Life

making copies of themselves, but prions achieve the same thing without any genetic baggage. They are just a bare, standalone protein.

Prions do their damage when they attack an important protein on neurons in the brains of their victims. They do their dirty work by altering the shape of that protein. And then they move on and do the same thing again and again. It is a molecular-level hack. The hacked protein is still the same string of amino acids that it was, but, after the hack, it is contorted into a nonfunctional shape. And that has literally mind-blowing consequences. As it turns out, a prion and its targeted protein are the same string of a couple hundred amino acids. The only difference between the prion and the nerve protein is how that string is twisted, folded, and wadded up. The prion is more compact. But, after a prion has pulled off its shape-shifting hack of the nerve protein, there is no difference between the two. They are both now twisted, folded, and wadded up in the same way. The neuron's protein has lost its ability to help coordinate brain functions, and, at the same time, it has been weaponized. It has become a prion, an infective agent.

A prion's highly compact form is resistant to heat (roasting or barbequing), digestive enzymes, and chemical treatments that would break up most proteins. In other words, it is a tough and persistent infectious agent (and, remember, not alive). Animal carcasses with prion-caused diseases must be handled and disposed of with great care to halt the diseases' spread. In one case, prions of chronic wasting disease (much like mad-cow except targeting members of the deer family) moved from the soil in which infected deer parts had been buried, into plants growing in that soil, and then into hamsters that ate the plants, causing the disease in the

Dead or Alive?

hamsters. In a Canadian study, monkeys contracted chronic wasting disease after being fed diseased deer meat.

It is not clear how species-specific the various prion diseases are. Mad-cow disease can infect humans. Scrapie infects both sheep and goats. Chronic wasting disease infects deer, reindeer, elk, and moose (and now hamsters and monkeys). Epidemiologists don't know whether prion-caused diseases can cross broader species barriers – as if from cow to human or deer to rodent and primate isn't broad enough. There's a lot yet to learn about prions, hopefully before they have forced themselves even more dramatically onto our attention. They are not alive, but they have some life-like features, and they can raise hell among the living.

From hacking to hijacking, but still not alive

Everyone has heard of viruses, but few are familiar with viroids. Viroids are infective agents in much the same way that viruses are, but they attack only plants. Probably even fewer people have heard of viroids than of prions. Diseases that make plants sick don't get as much press as ones that make holes in our brains.

Although we can assume they have been around for a long time, the first viroid-caused plant disease wasn't identified until 1971. That relatively recent discovery is a testimony both to the small size of viroids and to the difficulty of coming to grips with an entirely new kind of disease-causing agent. Plant pathologists reasonably expected the disease they are working with to be caused by a virus, a bacterium, or a fungus – not by something no one has ever heard of. Viroid diseases have now been found in several important crops to include potato, avocado, apple, and coconut. They

The Gyroscope of Life

are probably more widespread, but only plants with some economic value are likely to be tested for viroid infections. Viroid diseases are not fatal to the infected plants, but they often reduce or completely destroy the commercial value of a crop.

Viroids are of interest here because they can – just like prions – act like a living thing. But it is strictly an act. Just like a prion, a viroid is indisputably not alive. A viroid is a single, small RNA molecule. That's all. Yet, that small, naked piece of RNA can cause a disease in one plant, replicate itself by hijacking the RNA-making machinery of its host, and then move to a healthy plant and repeat the process. Viroids are not alive, but they behave in life-like ways. Viroids and prions remind us that life can be imitated by stuff that isn't alive.

Viruses: Victims of cellular bigotry?

In a consideration of disease-causing agents, viruses deserve more time and a higher billing than prions and viroids. Viruses cause many more diseases, and several viral diseases can be just as fatal as mad-cow or kuru. But the focus here is not on diseases; it's on the nature of certain disease-causing agents – deciding whether they are dead or alive. Prions and viroids definitely occur outside the life envelope, because they lack the complex components and biochemistry associated with living things. Each is a single, simple biochemical that exhibits some life-like properties. Viruses are more complex.

Viruses were discovered and beginning to be fairly well understood many years before prions and viroids were even suspected to exist. In fact, one of the well-established sub-

Dead or Alive?

disciplines of microbiology is virology. Having a biological field named after them, one might suppose viruses are considered alive, but they are not – at least not by many biologists. One indication that viruses aren't considered living is that, upon their discovery, they were not quickly given Latin binomial names, nor placed into any taxonomic grouping that showed a connection to plants, animals, or microbes. That is a telling snub. Virologists have gotten together in recent years and developed a scheme for classifying viruses that includes assigning names.

Viruses' major biological shortcoming, as far as many biologists are concerned, is that they are not cellular. According to most biologists, all living things are made of cells, which are characterized by being wrapped inside membranes. Viruses are membrane-less and therefore non-cellular. They are then, by definition, not alive. That bit of reasoning – living things are cellular, viruses aren't cellular, therefore viruses aren't living – is a product of a biological tenet formalized in the early 19th century, well before viruses were known. That tenet is the Cell Theory. We need to understand the basis of the Cell Theory to understand why many biologists consider viruses not alive.

Microscopes were invented in the 17th century, and, as the technology improved, naturalists were able to see things magnified 300 times or more. In the late-17th century, Dutch microscope maker and gentleman naturalist Anton van Leeuwenhoek saw what we now know from his drawings were single-celled animals and even some large bacteria. He was so secretive with his technology, though, that no one saw any more bacteria until the early 19th century. In fact, it wasn't until the mid-20th century, that someone was finally able to replicate the single-lens microscope van

The Gyroscope of Life

Leeuwenhoek had used.

As more and more microscopic observations were made of living things, it became clear that membrane-bounded cells were a constant. In the 1830s, naturalists proclaimed cells the basic unit of life. This became the Cell Theory, which says in part that life occurs only in cellular entities. But those naturalists may have painted biologists into a corner.

In the mid-20th century, an entirely new technology, electron microscopy, revealed viruses, which are much smaller than bacteria – and membrane-less. Without membranes, they were, by definition, acellular and therefore not alive. Accordingly, viruses were left on the outside of life looking in. Are they victims of a faulty theory because they were too small to be seen at the time the Cell Theory was being formulated? That would be ironic, given how many victims viruses have claimed via smallpox, pneumonia, Ebola, Black Death, AIDS, and rabies.

Everyone seems to agree that viruses live – or don't live – at the interface between biology and chemistry. Many virologists consider them more on the living side of that boundary. Some suggest viruses are evolving toward becoming cellular, living things. Others hypothesize viruses were once cellular and gave up some of life's key accoutrements for a simpler way of achieving the Prime Directive – to make copies of one's genes and pass them on. In any event, viruses are clearly good at that without needing all the trappings of life.

Cryptobiosis: Hidden life

Non-living entities like prions and viroids – and maybe viruses – give us pause, but some living things give us more reason for considering dead or alive a non-binary. That

Dead or Alive?

binarity was destroyed as we realized some organisms can totally suspend life for extended periods and then resume living. These beings can be clearly alive, then enter into a state where life is not possible, and then become clearly alive again. The general term for the intervening state of suspended animation is cryptobiosis, from the Greek for hidden and life. For those organisms that have evolved ways to do it, the most common reason for putting life on hold is lack of water.

 Water is essential for life as we know it. Life is thought to have arisen in the oceans. Water is a particularly favorable medium for life. It buffers against rapid temperature changes. Its buoyancy takes away problems that gravity can cause. It filters out UV rays, which can be harmful to living things. Water is also a solvent for biochemistry, and it is a reactant in numerous biochemical processes. Without water, there simply is no life – not as we know life. Plants and animals living on dry land can do so only because we are able to maintain an internal ocean in which our cells carry on life.

 However, evolution has fitted some organisms with strategies for remaining viable – meaning "able to live", not a synonym for alive – when their internal ocean evaporates. When dry, these organisms typically become quite fragile or brittle. But remarkably, when water becomes available again, they retain an ability to rehydrate, fire things back up, and resume life. Few higher life forms can survive significant dehydration. A well-hydrated human body is about 60% water, and most of that is inside our cells. A 10% loss of that internal ocean can have major impacts on body and brain performance. The loss of just a few percent more can be fatal. We would still have lots of water in us, but not enough to stay alive. But, for those organisms that have learned how to go

The Gyroscope of Life

with the flow and to take a time out when the flow stops, even complete desiccation is survivable. They can become dehydrated to the point that no liquid water is present and no sign of life occurs, and yet they remain able to resume life. That dry-down, shut-down strategy has been dubbed – from the Greek again – anhydrobiosis (without water, life).

A point in passing: biologists can be stodgy or nerdy types. When we discover new phenomena, we often seem obliged to dig into Greek or Latin to describe them. Hence words like anhydrobiosis (and, yes, autobiologic). Physicists are much cooler, coming up with creative and quirky names like Big Bang and quarks, whose official names include Up, Down, Strange, and Charm. I have introduced anhydrobiosis primarily to take exception to it. The name is quite misleading. Its Greek roots imply life continues in the absence of water. Life doesn't continue, as will be made clear in the next two sections. Some English phrases better convey what is really going on. "Suspended life" or "suspended animation" come pretty close.

Suspending life: A viable, non-living option for some organisms

The ability of some living things to dry to a crisp and then revive – sometimes decades or even millennia later – is fascinating. Many bacteria have forms that can suspend animation for centuries without water and then be revitalized (in the truest sense of that word) with a drop of water. Dehydrated baker's yeast, which is a fungus, is dry as dust and seemingly dead, but it comes back to life quickly after being reintroduced to water. Among microscopic life forms, some kinds of round worms can dehydrate to the point that they

Dead or Alive?

crumble if touched, and yet they can revive quickly when rehydrated (but not when crumbled). The round worms with this amazing ability to suspend animation include a quirky one that lives in beer-mug coasters. Presumably those tiny guys dry up happily there on the bar top and can be revived into a similar state the next time someone's beer sloshes out. A few macroscopic animals are also able to dry down and suspend life. Probably the most famous example of suspended animation among animals comes from brine shrimp whose dry eggs have been sold by the millions as "sea-monkeys".

Many practitioners of suspended animation come from the plant kingdom. One of my favorite examples is the resurrection fern. It is a smallish fern from the southeastern US, to include Appalachia, that often grows on trees or even on bare rocks. It draws nothing from its substrate – no nutrients, no water. Being on the side of a tree or rock, means that these ferns are periodically subject to severe drought – more so than the tree they might be living on. The tree's roots can continue to extract water from down in the soil, even after the soil seems pretty dry, and thus the tree maintains its internal ocean. Not so for a resurrection fern clinging to it.

I once had a pet resurrection fern, collected from an old elm tree in Tennessee. Her name was Polly, short for her Latin name, *Polypodium polypodioides*. She came to graduate school at Wake Forest with me, with the idea I might try to unlock some of her mysteries. She lived in a glass dish on the book shelf at my desk. There was no soil in the dish; just the fern with a short piece of stem, stubby "roots" (really just attachment devices), and five or six fronds. When well-watered, Polly would spread out pliable green fronds, soak

The Gyroscope of Life

up light, and carry on all vital fern functions.

But sometimes I would benignly neglect Polly. After a few weeks without water, she would lose 90 or 95% of her internal ocean. The signs of my neglect were a gradual downward curling of the fronds and then a decided tightening of the curls until the fronds became tightly coiled, shriveled, brittle, grayish, and decidedly dead looking. Polly could remain in that lifeless state for many weeks. Whenever I remembered to water her, however, Polly would spring back to life within a day or less.

As it turned out, I did not do any formal research on resurrection fern at Wake Forest. For my master's thesis, I worked instead with another drying-permissive form – seeds. Seeds provide the widest-spread examples of suspended life. Seeds of many plants can desiccate, or dry down, till only 5 or 10% of their full complement of water remains. In that desiccated state, they give no indication of being alive, and they can remain that way for years – or centuries in some cases – and spring back to life when they are rehydrated.

One of the best-documented cases of a dry seed reviving and germinating after many, many years is a 2,000-year-old Judean date palm seed. A cache of date palm seeds was unearthed in a dry location at an Israeli archeological site in the 1960s. The seeds were kept as museum pieces for another 40 years, but, in 2005, three of the seeds were planted, and one of them germinated. The resulting tree was named Methuselah, after the oldest man in the Bible. (Lazarus might have been more apt, except his story is not part of the Hebrew Bible.) Methuselah flowered in 2011 and is still alive as of this writing. The tree is unique not only for the amazing duration of the seed's viability, but also for the fact that it is the only Judean date palm in existence. The species has

Dead or Alive?

otherwise been extinct for about 800 years.

In another well-documented case, 32,000-year-old seeds produced new plants, but some midwifery was required. The seeds were from a small Russo-Arctic flowering plant. In 2007, more than half a million of the plant's dried fruits were found in ground-squirrel burrows buried under about 100 feet of permafrost. The ancient squirrels had damaged the seeds to the point that they would not germinate. However, in 2012, botanists were able to clone three dozen plants from 32,000-year-old fruit tissue. The resultant plants produced seeds that were quite germinable, so that the next generation did not take another 32 millennia. When compared to modern members of the same species, these Rip Van Winkle plants had slightly larger and more widely space petals, suggesting evolution had moved on while the seeds were spending all that time in suspended animation.

Then there is the well-publicized case of yeast – yeast cells, not seeds; yeast don't produce seeds – being brought back to life after being encased in amber for 45 million years. One can buy beer brewed with the ancient yeast. Or so the claim goes. Two scientific reservations revolve around this claim. No other work suggests that genes in naturally preserved materials can survive in a functional form for more than 1 million years. Furthermore, no independent lab has been able to regenerate live yeast from millions-of-years-old amber samples. The jury may still be out, but the scientific evidence seems to weigh heavily against the claims, although the beer gets good reviews.

In dry seeds, life isn't just hidden; it's absent

Whoever coined the term anhydrobiosis (without water,

The Gyroscope of Life

life) apparently did not have a good grasp of what is going on – not going on, actually – in dehydrated seeds, resurrection ferns, yeast, brine shrimp, and roundworms. These drought-tolerant organisms can dry down till only 5% of their weight is water. Biophysicists tell us that residual water is not in a liquid state. Rather, it is tightly attached to cellular surfaces in ice-like layers. How do we know any water is still there? Heat it up to about 150°F. The tightly bound water evaporates at those temperatures, and the weight of the organism drops by a final 5%. Seeds held in this bone-dry state for extended periods can still be revived. Put these facts together, and we can say with great confidence that liquid water is not required in seeds adapted to drying down.

Still, some argue that dry seeds continue to carry on life. Their argument is driven by vitalist and/or binary thinking. They reason that dry seeds must be alive, since rehydration proves they were not dead. The seeds must either be retaining a life force (a vitalist's argument) or continuing to carry on life processes (a "binarist's" argument), even while dried to a crisp. Since biologists cannot detect or measure a life force, most reject the vitalist argument. But what about the argument that life continues in dry seeds, albeit at a low rate?

Respiration, which is an oxygen-consuming metabolic process, is often cited as a sign of life. If we put dry seeds into a device that measures oxygen levels, we can detect minute amounts of oxygen being consumed. To some, that serves as proof (something that science doesn't provide, actually) that the seeds are respiring at a low rate, and, if they are respiring, they must be alive. Or so the illogic goes. But we can put iron filings in the same device and detect similar amounts of oxygen consumption. Tantalizingly, the amount of oxygen consumed increases if we add water. A sign of life?

Dead or Alive?

Hardly. When iron oxidizes and forms iron oxide, or rust, oxygen is consumed, and water accelerates that oxidation.

So, what is the explanation for the uptake of oxygen by dry seeds and other anhydrobiotic organisms? It is quite analogous to rusting. Dry seeds experience deteriorative rust-like processes, but iron is not being oxidized. The most likely candidates for oxygen attack are lipids (generally, oily substances). Lipids are particularly susceptible to oxidation. When exposed to the oxygen in air, vegetable oil will become rancid, a result of lipid oxidation. The same is true for lipid-rich nuts and seeds: peanuts, pecans, sunflower seeds, etc. In all seeds, lipids are major components of cell membranes. In some seeds and nuts, lipids are also a major energy storage form. So, an undesirable oxidation of lipids – not respiration – can account for the uptake of oxygen in dry seeds.

Here is a final nail in the respiration-must-be-happening-to-keep-the-seeds-alive coffin. Dry seeds can be stored in oxygen-free environments for long periods and not be any the worse for it. Indeed, they are the better for it. The viability of many seeds is best maintained in environments that are dry, cold (even down to absolute zero), and oxygen-free – all factors that prevent respiration.

Bottom line: respiration is unnecessary – not to mention impossible – in fully dehydrated seeds, resurrection ferns, sea monkeys, yeast, etc. It is not possible for life's chemistry – including respiration – to happen without liquid water. Without water, life ceases. The Greek-loving originator of "anhydrobiosis" seems guilty of assuming a dead-or-alive binarity. He or she knew dry anhydrobiotic beings could not be dead, because they are quite capable of living if provided water. So he or she presumed they stayed alive. They do not. They cannot. Anhydrobiosis be damned. Life is suspended in the dry state for these crispy creatures. Dead or alive simply is not binary for them.

Aging while time stands still

There is a definite down side to being neither alive nor dead. Dry seeds are in a state of suspended animation; the world goes right on while time stands still for them. That is a problem – a problem that only gets bigger the longer a seed stays off line. Lipids are attacked by oxygen. Other complex, energy-rich chemicals, such as proteins and DNA, slip toward greater entropy. Bugs and ground squirrels gnaw. Dry, brittle seeds get banged around and cracked or broken, especially if they are being harvested and handled mechanically. Sometimes seeds pick up enough moisture to let fungi or bacteria grow but not enough water to let the seeds' defense and repair machinery get going. It's the 2nd Law of Thermodynamics in spades; shit is happening. Entropy is increasing. And here is the rub: dry seeds can do nothing about any of these threats until they rehydrate and fire back up their metabolic machinery. They may have had their shit together as they were drying out, but they can do nothing to keep it together while life is suspended.

The accumulation of damage – those disordering, entropy-raising events – in dry seeds affects seed performance upon rehydration and reanimation. Over time in suspended animation, seeds become less able to produce healthy seedlings. This decline in eventual performance that develops while seeds are dormant or in storage is called seed aging. It is caused by random, entropic events that occur at a molecular, cellular, or whole-seed level while the seeds are dehydrated, and it reduces their ability to resume life

Dead or Alive?

upon rehydration.

A life-or-death crisis for aged seeds may come during rehydration. Oxygen-damaged membranes may not be able to keep everything in the right place as water rushes back in. Studies have shown that gradual rehydration, where aged seeds' internal oceans are refilled slowly, can make a difference in whether those seeds end up being dead (nonviable) or alive (viable). Shades of Schrödinger's cat! (See next section for more on this famous feline.)

Aging also happens in organisms that are fully hydrated. Plants and animals whose internal oceans are at full pool and that are fully alive are still susceptible to the slings and arrows of entropy. Over time, our bodies collect random damage from injury, disease, toxins, and the inevitable entropic wear and tear on our constituent parts. But we are at a decided advantage relative to dehydrated seeds – as well as resurrection ferns, sea monkeys, and other organisms in a state of suspended animation. We have defense and repair mechanisms working for us that can protect us from or repair entropic damage. Dry seeds, ferns, or round worms do not. By some point, they accumulate so much damage that they will no longer be alive when rehydrated. At that point, they can be declared nonviable, or dead. Until that point, they are viable, but neither dead nor alive. This conundrum seems to call for a timeout to consider just what life is. What does it take to be alive?

A thought experiment about life (and death)

Thought experiments are intellectual exercises, flights of imagination in which scientists, philosophers, and mathematicians wrestle with concepts that may not lend themselves

The Gyroscope of Life

to physical experimentation. Some thought experiments are famous. Schrödinger's cat, which is simultaneously dead and alive, helped bring light to one of quantum mechanics' more intriguing properties. Maxwell's demon is an imaginary mechanism that can violate the 2nd Law of Thermodynamics, allowing heat to accumulate rather than dissipate. In one of Einstein's thought experiments, he chased a beam of light. It was useful to him in his development of the theory of special relativity.

Galileo, one of the more famous natural philosophers, used a thought experiment to examine the effects of gravity. Since Aristotle, and because of Aristotle, people believed a heavy object would fall faster than a light one. In a famous and perhaps apocryphal story, Galileo climbed a tower in Pisa (yeah, that one) and proved that heavy and light objects fall at the same pace. But a well-documented thought experiment by Galileo demonstrated both the fallacy of Aristotelean thinking about gravity and the power of thought experiments. Galileo posed his thought experiment this way: Here are two spheres of different weight. If, indeed, they fall at different speeds, what will happen if the two are tied together? The lighter sphere should lag behind and slow the descent of the heavier rock. But the two collectively now represent a single, heavier object that should fall faster than either alone. The incongruity of these two contradictory outcomes suggested Aristotle's thinking was faulty.

For whatever reason (stodginess?), biologists seem less likely to use thought experiments. Few have stood the test of time anyhow. Here comes one that will never rank up there with chasing sunbeams, but it will hopefully provide some insight into what it takes to be alive. We have here before us two objects. The one on the right is a dry seed taken out

Dead or Alive?

of a packet bought at a garden store. The object on the left looks exactly like the one on the right. It seems to be another dry seed. But this object was not made biologically. Instead, it was produced by some accomplished technicians. Those technicians were so skilled that their fake seed is an exact copy of the real seed: exact down to the last detail of every molecule and atom in the real seed. We will eventually ask how those two objects will behave if we add water. For a moment, let's dissect the proposition.

For multiple reasons, this must be a thought experiment, not one that could actually be performed. The first problem is that no one knows exactly how every molecule and atom in a seed (or any other life form) is arranged. We have only general ideas at best about some details. A second reason the thought experiment can't be taken into the lab springs from another limitation. Even if our hypothetical technicians had a blueprint that showed the location of every atom and molecule in a seed, they don't have tools to assemble atoms and molecules into those locations. And, building on the absurdity of doing this as other than a thought experiment: even if we had a blueprint and the tools, it would take an awfully long time to properly assemble the trillions of atoms in a seed. Walt Whitman's line comes to mind: "…a leaf of grass is no less than the journey-work of the stars". (Whitman was not aware of the creation of elements in exploding stars, but that line from "Leaves of Grass" seems prescient.)

So, back to our thought experiment. Consider again the two objects before us. The one on the right is a normal, naturally produced seed. The one on the left was built to be exactly like the real seed, down to the nth detail. Now, let's take the two objects and put them in a warm, moist soil. What will happen? We can easily predict that the seed from

The Gyroscope of Life

the garden store will germinate and produce a healthy seedling. But what about the facsimile?

Our thought experiment is something of a Rorschach test. Vitalists are likely to predict that the ersatz propagule will do nothing, because the vital force cannot be synthesized. Holists might be inclined to say the lab-assembled seed would germinate but might still argue that a real seed would somehow equal more than the sum of the parts of the one cobbled together. Thoroughgoing and irredeemable reductionists, such as myself, have no choice but to predict that the facsimile will behave like the real McCoy. We believe life ultimately is the result of a precise organization or arrangement of matter. If we arrange non-living matter into a configuration that exactly duplicates living matter, the former will behave like the latter; they will both be alive. Unless life is ultimately a mystical phenomenon, a perfectly faux seed should behave the same way a real one does. Each will have the attributes we associate with life.

Defining life

The previous sentence raises a rather momentous question: what are the attributes of life – the characteristics that cause us to describe some collections of matter as alive? What properties are unique to living things? Maybe we should have broached this line of questioning sooner. It is screaming for attention now. To a reductionist, such as myself, cells are the logical place to look for the answer to what constitutes or identifies living matter. Cells are, after all, the irreducible units of life. But even reductionists describe organisms made of trillions of cells as alive also. What list of attributes can be used to ascribe life to a microscopic bacteri-

Dead or Alive?

um as well as a 100-foot long, 200-ton blue whale? There are just three items on my list. Anything that possesses all three is alive. And, as it turns out, we have already identified those three items. We just haven't described them as life-defining terms.

Since the 19th century, one well-agreed-upon characteristic of life has been cellularity. Everything alive is a unit – or made of units – surrounded by membranes. (I see your hands going up, virologists. Will you be okay if I say, "Everything that non-virologists recognize as alive…"?) For most of us, all living things are cellular. That was codified in the previously discussed Cell Theory.

Also already covered at some length is this defining characteristic of living things: as chemically and organizationally complex as they are, they can keep their entropy low. In my view, this is life's most remarkable property because the 2nd Law of Thermodynamics assures us that collections of complex, energy-rich matter cannot persist indefinitely. Living things seemingly defy the 2nd Law – while obeying HR#2 – by continually obtaining energy and matter from their environment to maintain their critical living order – as well as grow. (Minerals are said to grow, but "form" is a better term. Mineral crystals form – in complete compliance with thermodynamic laws – as geologic and hydrologic forces act on various elements in the Earth's crust and on the surface. Minerals do not grow by dint of their own efforts or devices. Living things do.)

Finally, living things are able to obey HR#3 – to reproduce. Technically, producing more of one's self is not an absolute requirement for life. The role of some specialized cells is just to maintain themselves and not divide, and sterile plants and animals are nonetheless alive. But every living

The Gyroscope of Life

thing makes it into the land of the living only after being produced by another living thing.

So, the phenomenon we call life is a collection of processes carried out by living things such that they are self-maintaining and self-perpetuating. Accordingly, any entity with just three attributes – cellular, able to keep its entropy low, and able to reproduce – is alive.

The Gyroscope of Life: Cellular level

Here comes our metaphorical, autobiological, and eponymous way to consider the dynamic between matter and energy in living cells. Think of life in a cell as being stabilized and maintained gyroscopically. Gyroscopes are the top-like toys you may have played with – a spinning disk (called a flywheel or rotor) mounted in a frame, which may be mounted in concentric gimbals attached at pivot points. Spin the flywheel, and it will remain quite stable in its orientation when the surrounding gimbals are rotated in any direction. Besides being a fascinating toy, gyroscopes serve serious purposes. In airplanes, they are the heart of instruments that keep track of the horizon. In tunnel digging, they keep the boring machines on track. In space telescopes, such as Hubble, gyroscopes keep the optics pointed steadily at things far outside the solar system while the telescope is moving through its Earth orbit and the Earth is moving through its solar orbit. The physical principle behind a gyroscope is angular momentum; once the rotor is set spinning, it will tend to stay in that orientation. Flywheels that are heavier, better balanced, and spinning faster provide greater stability.

Something of a similar nature – a metaphorical Gyroscope of Life – stabilizes, or preserves, life in cells. Nothing

Dead or Alive?

is actually spinning, but processes are in motion that keep the cell stable and pointed in the right direction, toward maintenance, growth, and reproduction. The flywheel of this metaphorical gyroscope is kept spinning by energy obtained from the environment. For cells in green plants, the energy to spin the rotor is sunlight, which can be converted into chemical energy. For the cells of most non-green organisms, the energy that spins the flywheel comes as chemical energy passing through food chains. In either case, a continual supply of energy can keep the flywheel spinning at high rpms, which bodes well for the life of the cell.

The metaphoric flywheel, or rotor, is represented by the entropy-defying complexity and precise organization of living cells. Life is the result of a highly ordered heterogeneity, as oxymoronic as that phrase sounds. Life happens when all the critical subcellular components are present and in good working order. Life has a place for everything, and everything is in its place when life is going well. When order at the cellular level is optimal, the flywheel of the cell's Gyroscope is balanced. If some of that order is lost, the flywheel becomes unbalanced. The cell can continue to pump energy into keeping the rotor spinning, but it is going to be a wobbly ride. Toxins, viruses, radiation, injury, etc. can cause disruption of cellular order such that the flywheel is made less balanced. If too big a chunk is knocked off (perhaps something messes up membranes' control of what gets into and out of the cell, or UV radiation causes genetic disruptions), the rotor can be fatally destabilized, and the cell will die.

The good news is that living flywheels are self-repairing to some degree. Cells can often detect when entropy creeps up, and they can repair the problems. Attacks by oxygen on membrane lipids occur in hydrated tissues, and they can be repaired when the metabolic machinery is up and running. Plant and animal cells can sometimes detect and repair damaged DNA. As molecules of a protein become degraded by wear and tear, new ones are produced to replace the faulty ones. All these actions serve to repair damage and lower entropy back toward optimal levels to keep the flywheel balanced and spinning smoothly.

Let's consider again the status of the cells in organisms that can dry to a crisp and suspend animation. No metaphorical flywheels are turning in dehydrated cells. As a cell dries, its metabolism – its biochemistry, the chemistry of life – slows down and then stops. The molecules that make up the figurative rotor are still there, but it cannot spin. Water is, in essence, a lubricant for life. Without it, the rotor comes to a halt. If all goes well, it will resume spinning as soon as metabolism is restored. But, if entropy has nibbled on the flywheel too much while the cell is dry, the flywheel may become badly unbalanced. When water is restored and cellular metabolism tries to fire back up, the beat-up gyroscope sometimes can no longer do its job, and the cell's life will sputter out.

The Gyroscope of Life: Organism level

Cells are the basic unit of life, and we've just considered autobiologically how we might define and metaphorically explain what it takes to be alive or dead at a cellular level. However, we also describe as dead or alive organisms that consist of billions or trillions of cells. What defines and explains life in an entity that is a collective of trillions of living entities? As an opener, let's consider that a Gyroscope of Life operates at an organism level also. The energy sources that

Dead or Alive?

spin the rotor remain the same as for the cellular gyroscopes – sunlight and biochemicals passing through food chains. The nature of the flywheel shifts away from being a precise organization of matter in cells into being the systems and processes that help maintain and reproduce the collective of cells. The cellular Gyroscopes remain essential. Each cell must do its part and maintain its order. But new, more complex levels of coordination and interaction must be regulated also.

Complex, multicellular organisms like higher plants and animals are often described as being composed of multiple organs or organ systems. Animals have specialized cells that make up circulatory, respiratory, digestive, skeletal, muscular, reproductive, etc. systems, each with critical functions to perform. Plants have root, stem, leaf, and flower organ systems. These organ systems collectively constitute the organism, and they each must operate at peak efficiency to allow the organism to prosper. They can, for our purposes, be thought of as the organism's flywheel. When they are all in balance and operating smoothly, the organism can carry on life.

Mechanisms that maintain or regulate the stability of critical functions in multicellular organisms are often described as homeostatic, from the Greek for (more or less) "staying the same". The human body has an internal thermostat that regulates body temperature. It can trigger sweating if our core temperature gets too high. When we get cold, homeostatic mechanisms can increase our metabolic rate to generate more heat by converting chemical energy into heat energy. If that doesn't do the job, we will begin to shiver, because the contraction of muscles generates heat as a by-product. Blood sugar levels are another factor monitored and

The Gyroscope of Life

regulated by homeostatic sensors and responses. Diabetes occurs when the homeostatic system for blood sugar falters. The result can be either too much or too little sugar in the blood and a destabilization of both cellular and organismal Gyroscopes of Life.

Diseases and toxins often disrupt homeostatic processes. I have advanced prostate cancer. Rogue prostate cells are already present in lymph nodes and bones, and more systems will likely become compromised as the disease progresses. As I was rewriting this chapter, I underwent chemotherapy, and my body clearly let me know that homeostatic processes were set teetering by that toxin. The good news is that round of chemo is behind me, and I can tell the flywheel is repairing itself and beginning to pick up speed again. We haven't kicked the cancer's ass, but we have kicked the can down the road a bit, and my Gyroscope of Life is at least temporarily spinning more smoothly.

Plants exhibit homeostasis too – mechanisms that collectively constitute their organism-level Gyroscope of Life. One of those mechanisms helps to maintain a balance between roots and shoots. As a generalization, a plant will have as much of a root system belowground as it has stems and leaves aboveground. Above- and belowground parts need to be in balance. If shoots' volume exceeds roots', the roots won't be able to keep up with the shoots' demand for water and minerals. Likewise, if shoots are lost to disease or injury, the leaves and stems won't be able to provide all the sugars and other complex materials that roots need to do their job. How does the plant maintain a suitable root-to-shoot balance? Root systems make a hormone that moves upward and triggers shoot growth. The shoots make a different hormone that moves downward and triggers root growth. If the root

Dead or Alive?

system is damaged, growth of shoots will be slowed because they are no longer getting a strong "grow" signal, and vice versa. As long as that homeostatic system is operating, the root-to-shoot ratio will be kept in or returned to a balance, and the organismal Gyroscope can continue to spin smoothly.

Maintaining an internal ocean is another crucial requirement for plants and animals living on dry land. Plants have multiple homeostatic ways to keep themselves from becoming dehydrated. When many plants experience drought, they produce a chemical that causes their leaves to fall off. It's a drastic solution, but it can keep the plant alive until water again becomes available. It's a solution often seen in Mediterranean environments where there are distinct wet and dry seasons. Plants produce leaves during the rainy season and then homeostatically shed them when it gets dry.

One of plants' most sophisticated homeostatic mechanisms for maintaining that internal ocean is seen in the role that stomata play. Stomata are microscopic holes in a leaf's surface. That surface serves as a dike between the plant's internal sea and the atmosphere's external desert. The stomata's primary role is to let carbon dioxide come inside leaves for photosynthesis. But there is a rub; at the same time carbon dioxide gets in, water vapor gets out. Those stomata, and there are thousands per square inch, are holes in the dike.

But stomata are not always a hole. They can be fully opened or fully closed or somewhere in between. Regulating the size of each stomatal opening are two sausage-shaped guard cells. Put your flattened hands together palm to palm. Now flex so that only your finger tips and the heels of your palms still touch. Now flatten them again. That's how guard cells open and close stomata. But how do guard cells know when and how much to flex? Their relaxed state – no opening – is the norm. At night, the stomata are tightly shut, keeping water inside the leaf. When the sun comes up, both light and falling internal carbon dioxide levels trigger stomata to open. The dike is suddenly full of holes. The stomata are inviting disaster. But multiple homeostatic mechanisms are monitoring and responding to leaf hydration levels. When guard cells lose too much water, they relax so that they come to lie alongside one another again and the stomata close. That has a downside as far as photosynthesis is concerned, but the downside of becoming too dehydrated is much greater. Homeostasis keeps the Gyroscope spinning.

Many other processes in plants and animals are under homeostatic control, where the integrity and stability of the organisms are maintained or stabilized gyroscope-style. But what happens if those stabilizing homeostatic systems become compromised by environmental hazards, toxins, aging, disease, or other disrupting influences? Each homeostatic system contributes its part to the flywheel. When any of those systems is damaged or weakened, the flywheel may become unbalanced. The collapse of any homeostatic process can lead to complete destabilization of the rotor and a Gyroscope that stops spinning, i.e., death.

Some final words on finality

For unicellular organisms, a single cell is the only unit of life. When that cell dies, the organism is dead. But, what about death in multicellular organism? Each cell in a multicellular organism is a living unit, and each cell's life can continue for varying periods after some or many of the other cells have died. An organism can be visibly dead in terms of

Dead or Alive?

its overall status, and yet many of its cells could still be alive – at least for a while. We're not talking about the living dead here. We're just trying to get a hand on what death is. It's a fuzzy area.

For human beings, clinical death occurs, or is pronounced, when heart beat and breathing cease or when vital brain functions have ceased – brain death. But that doesn't mean everything in the body is dead – not instantly. With good medical management, organs may be removed from a recently deceased person and placed into another person many hours later. Donated organs must remain alive in this transition. The greatest threat to their wellbeing is running out of the oxygen needed to keep cellular metabolism going and cellular Gyroscopes spinning. Oxygen is normally delivered by the actions of the respiratory and circulatory systems. When a donor's respiratory and circulatory systems fail, all the other systems began quickly to crash. But machine-assisted blood circulation and breathing keeps their cells supplied with oxygen for respiration. That keeps the organ's cellular Gyroscopes spinning, and those still living cells keep on keeping the organ's entropy at bay.

The death of higher plants can be much harder to define or delineate. There is no legal or clinical definition for the death of a redwood tree or a resurrection fern. In fact, trees are largely made up of dead cells even when they are perfectly healthy. The wood that constitutes 99+% of the mass of redwoods and most other trees is made of hollow, dead cells. That is their functional state. They have living components during their development, but the living portion of wood cells must die before sapwood can do its crucial job of carrying water up the tree. The dead woody cells produce microplumbing for water movement.

The Gyroscope of Life

In the non-woody portions of plants, most of the living cells can die, and yet the plant can live on for years or centuries. Mark Twain once wrote, "The report of my death was an exaggeration." That was later – after his actual death – transformed by a biographer into, "The reports of my death have been greatly exaggerated." Plants might say the same thing, if they could say anything. Plants can dwindle to just a few living cells and from that remnant regenerate a whole, new plant. In the laboratory, it is possible to regenerate a whole plant from a single cell. That process is generically called cloning. It has been used with great success commercially – and also in resurrecting 32,000-year-old seeds. (Animals have also been cloned in the laboratory, but the technical and biological hurdles are considerably greater.)

So, death is fuzzier in multicellular plants than in multicellular animals. But, even for the animals, death is not an organism-level event – except where defined for legal purposes. It's a cell-by-cell process in which entropy finally wins the field; in which the cellular Gyroscopes of Life teeter and rattle to a halt one by one.

7

Plant or Animal?

Natural bodies are divided into three kingdoms of nature: namely, the mineral, vegetable, and animal kingdoms. Minerals grow. Plants grow and live. Animals, grow, live, and have feeling.

Carl Linnaeus (1707 – 1778)

In the beginning

In the biblical creation account, the only living things made were either plants or animals. That scripturally established biological binarity – plant or animal – fit nicely with everyone's observations from prehistoric times right on up until about the 18th century. Throughout the history of natural history, living things were sorted into just two bins. If it was alive, it was either a plant or an animal. Plants were the sedentary or floating green things plus the sedentary or floating non-green mushrooms and molds. Animals were the slithering, crawling, walking, flying, or swimming things that typically responded to being poked. With the appearance of primitive microscopes in the 17th century, previously invisible things were discovered living in drops of water. If they were green, they were plants. If they moved, they were animals.

The development of higher resolution microscopes by the late-17th century revealed even smaller things in drops of water. They didn't swim around, and they weren't green, but they grew and made more of themselves. Were they plants or animals? As eventually became clear, they were bacteria, good old Life 1.1, neither plants nor animals. That is the problem with thinking dichotomously when one is tackling a non-binary subject. Just as matter doesn't have to be either dead or alive, living things don't have to be either plant or animal. Or so it was eventually – maybe grudgingly – determined. How we came to recognize that diversity makes an interesting story.

Finding living things and naming them

Naturalists, the forebears of biologists, loved finding and naming things. Biologists still do. It is a useful compulsion. It's even biblical. In the creation account in Genesis 2, Adam got to name each of the animals brought to him after they were created. The name for naming things is taxonomy, from the Greek for arrangement and method. Synonyms for taxonomy, or closely related to it, are classification and nomenclature. Adam was a nomenclaturist.

After Adam, the next-most-famous taxonomist is Carl Linnaeus, an 18th century Swedish physician and gentleman naturalist. He developed a binomial (double-name) Latinized nomenclatural system, along with rules for how it was to be used. Linnaeus is credited with naming over 12,000 species himself. This was in the days when naturalists were assiduously collecting, curating, and celebrating newly discovered "natural bodies" (see Linnaeus's opening quote). Putting a unique name to each newly discovered organism helped bring order to an exuberant field. Putting a unique Latin name to familiar species was important also. Common names – whether in Swedish, English, German, French, etc. – for various plants and animals varied from place to

Plant or Animal?

place, and the same common name was sometimes used for completely different organisms. Linnaeus's contribution to the classification of living things has been a boon to biologists. It provides every species its own two-word Latin name. That allows biologists to put a definitive name on any organism they might study and to distinguish their findings from those on any other species.

Furthermore, Linnaeus cultivated ideas that may have helped taxonomists move closer to understanding the evolutionary connections between living things. (Darwin wouldn't open that can of worms until the middle of the 19th century.) Besides busily assigning binomial names to the stream of exotic animals and plants being sent to him, Linnaeus looked for ways to systematically organize them. He created classification categories named kingdom, order, and class plus the genus and species ranks implicit in a binomial name. He did not understand that those groups sometimes revealed evolutionary connections. They were merely ways of bringing order to the chaos that living things might otherwise represent. Over the next couple of centuries, his taxonomic scheme was expanded into a seven-tiered system with the bottom rank (species) being the most exclusive and the top rank (kingdom) being the most inclusive. Some of Linnaeus's attempts at organizing species into the intermediate taxonomic groupings went a bit awry, but his uppermost category—kingdom – was solid. Being a product of his time, he identified two kingdoms of living forms: the animals (Animalia) and the plants (Plantae). A nice, dichotomous binarity.

In the last 250 years, perhaps 1.5 to 1.9 million species have been described and named using Linnaeus's method. That "perhaps" is a tip of the hat to the fact that no single

The Gyroscope of Life

governing body or authority registers, certifies, and compiles the Latin names generated across years, across taxonomic kingdoms, and across the globe. By some estimates, new binomials are being created and published all over the world at the rate of 50 or so each day. Capturing all of those reports, validating them, and entering them into an official global registry would be a Herculean task, and no one is paying Hercules – or anyone else – to do it. That estimate of 1.5 to 1.9 million named species might be what we called a WAG (wild-assed guess) in my military days.

1.5 to 1.9 million named species is impressive at face value anyway, but one fairly recent estimate of the likely number of extant species of higher-life forms – plants, animals, and fungi – came to 8.7 million. That estimate is based partially on knowledge that many of the world's richest biological zones have been only lightly canvassed for biota. Using data from the portions of rain forests, marine environments, etc. that have been canvassed extensively, taxonomists project that most species remain to be named.

Botanists have done a pretty good job with finding and naming plants. Close to 90% of an estimated 0.4 million plant species have been described and named. Of the 6.8 million animal species zoologists may have to work with, about 90% are insects. As it turns out, entomological taxonomists seem to have a lot of collecting, describing, and naming to do; 80% of a putative 6.1 million insect species have yet to be named. Mycologists are doing no better taxonomically. In fact, maybe 90% of an estimated 1.5 million fungi remain unnamed. There could be some job security in taxonomy. Oh, and then there are all those extinct species that need names…

Plant or Animal?

Finding order and a system (a long sidebar)

If Earth's estimated 8.7 million species of animals, plants, and fungi are simply lumped together, we get a metaphorical mincemeat pie. I suspect many folks are like me; even if we know generally what mincemeat is, we have no idea what's really in it. Mincemeat pie is an Appalachian favorite that harks back to the British Isles – from whence most mountain people trace their lineages. Mincemeat is famous – or infamous – for having many and varied ingredients, to include fruit, spices, liquors (such as moonshine), suet, and actual minced meat (such as wild venison). When I was growing up, mincemeat pies usually appeared around Thanksgiving and were heavily flavored with sweet fruits – and sometimes brandy at non-teetotalers' tables – as a dessert, rather than a savory main dish.

Taxonomists can name the ingredients (species) in our metaphorical living mincemeat, but, just as with real mincemeat, the real mystery comes in figuring out how those parts fit together – how they come to make up a pie. Helping us understand how the living pie is assembled is the job of another order of the taxonomic priesthood – systematics. Systematists are generally considered the high priests of taxonomy. They look for patterns of shared features, commonalities that identify species with a shared evolutionary history. They are seeking the metaphorical equivalents of fruit, spice, alcoholic, suet, and meat categories in mincemeat, and they tend to have strong opinions about how to bring order to what otherwise seems a living jumble. But, let's get away from some weird-tasting pie and use another metaphor to describe the importance – and difficulties – of systematics.

If my old car is having a problem and I call a mechanic,

The Gyroscope of Life

he or she is going to want to know what I drive. It won't help if I say, "So, it's a vehicle." That would be about as informative as saying, "So, it's red." There are 275 million vehicles on US highways. The mechanic needs to quickly eliminate from consideration most of the 275 million possibilities and get to the vehicle that is our concern. Let's think of a good, systematic way to do that. We could initially divide all 275 million vehicles into a few groups based on manufacturer (Ford, General Motors, Chrysler, etc.). Then, within the GM group, for example, we could sort by brand name (Chevrolet, Buick, GMC, etc.). Then within the Chevrolet brand, we could sort by model (Impala, Volt, Camaro, etc.). Then by year of manufacture. So, instead of just saying my car is a vehicle, I might say, "It's a 1955 Chevy Corvette". There aren't a lot of that species of car around. That shorthand way of describing – or classifying – the vehicle has great informative value.

Hang in here, please. I'm going to expand on the notion of classifying things for reasons that will become apparent shortly. We could create a similar classification scheme for airplanes or boats. We could create a nomenclature that systematizes transportation modes on land (from pogo sticks to bullet trains), in the air (from parachutes to rockets), or on water (from inner-tubes to submarines). If we got really ambitious, we could slam all three of those into a single scheme that classifies all forms of land, air, and water transportation. The first division in such a grand classification scheme would likely be between land, water, and air modes. My fictitious 'Vette would be in the land category and could be further classified as a wheeled vehicle, then an on-road vehicle, and then a car.

Taxonomists go through a similar process, applying their

Plant or Animal?

organizational skills to living things. They begin by sorting life forms into a few major categories (the land, water, and air equivalent). Then each of those more inclusive groups is divided into subcategories (wheeled, tracked, sliding equivalent), and sub-subcategories (on-road, off-road, on-tracks equivalent), and so on. With a system for categorizing, or classifying, living things, order appears out of the chaos, and these categories draw attention to consideration of how those traits came to be shared by so many species. Systemacists earn their salary by creating groups that are based on evolutionary kinships. Ideally, every species within a well-developed taxonomic grouping can trace its lineage back to an organism that first evolved some unique feature. That Adam/Eve individual is an ancestor of all the organisms that now bear that feature and that constitute that taxonomic grouping. Systematics' ultimate goal is to organize all living things into what is often described as a "Tree of Life". These are family trees, showing connections between relatives. Systematists look for kinships, i.e., evolutionary connections, between worms and frogs, between flowering plants and ferns, and so forth. To date, the high priests have developed no consensus for what the Tree of Life looks like, but a general shape is coming into focus.

In what follows, I am going to deal primarily with kingdoms and how they have been staked out by systematists. Looking at kingdoms is where we begin to understand that life is not made up of just animals and plants. They are, indeed, each in their own kingdom, but they share that honor with other groups.

Making top-level decisions: Busting one binary and avoiding another

The Gyroscope of Life

Systematists have been slow to come to agreement about how the pie of life should be sliced – not a mincemeat pie now, but a pie chart that would arrange living things into their highest-, or kingdom-, level categories. Have you picked up on a pattern here? Biologists often disagree. Systematists may be among the most likely to disagree, and one of the more contentious matters in systematics over the years has been where to make the first few kingdom-identifying cuts in the pie of life. Those wedges from a pie chart would correspond to the land, sea, and air categories in our would-be classification of modes of transportation. Those first few pieces need to be cut with care. So must the finer and finer slices, but, if the first delineations are not true-to-life, all the rest will get messier. And systematists have argued mightily over where those first few cuts should go, how many slices (kingdoms) there should be, and even about expanding the Linnaean system.

Linnaeus's 18th century classification system cut naturalists' pie into thirds. Being a thoroughgoing naturalist, he identified three kingdoms to which "natural bodies" could be assigned: animals, plants, and minerals. Yes, minerals, using the term in just the way we do today. After all, minerals were a legitimate subject for natural historians. Linnaeus did not consider minerals to be alive, but he thought they could be put onto the taxonomic rack that he had created. Mineralogists eventually found that much too tortuous and torturous. By the middle of the 19th century, Linnaeus's third kingdom had died a merciful death. That's not the best metaphor perhaps, since minerals were not alive to begin with. In any event, the proto-biology side of natural history still found itself with just two kingdoms – animals and plants. A binarity.

Plant or Animal?

In the second half of the 19th century, biologists' binary dam began to develop cracks. Systematists began to struggle with classifying single-celled algae and amoeba as plants and animals. Such single-celled organisms looked so different from other plants and animals that a British systematist proposed in the 1860s a new kingdom for them – the Protoctista. That name was eventually shortened to Protista, and the new kingdom was reaffirmed in other systematists' schemes. But heavy hangs the head that wears the crown. Since their coronation, the Protista have been periodically dethroned and re-instated by various 20th- and 21st-century systematists.

Another kingdom was in the making, even as systematists wavered about the Protista. As bacteria came more and more into focus – both microscopically and as a topic of systematic discussion – they were elevated into higher and higher taxonomic ranks – initially within the plant kingdom. In 1866, a German systematist put them into their own phylum, Monera (in the newly formed Protista kingdom). The name reflected bacteria's single-cell nature, "moneres" being Greek for solitary. In 1925, a French marine biologist, suggested the monera/bacteria were so different from other living things that they should be in their own kingdom, and bacteria have not dropped below that taxonomic rank in the estimation of most systematists since. The binarity was shattered.

When I was taking high-school biology under Coach P in 1960, it was a three-kingdom world: animals, plants, and bacteria. Throughout my undergraduate years, life continued to be encompassed in three or four kingdoms (with Protista being the odd man out). But, just after I finished my PhD work, in 1977, bacterial systematists split the bacteria into

The Gyroscope of Life

two kingdoms, the Bacteria (or Eubacteria) and the Archaea (or Archaebacteria). Then, around the dawn of the 21st century, yeasts, mushrooms, and molds were pulled out of the plant kingdom and given their own kingdom, Fungi. The decision to put fungi into a separate kingdom came easily enough after they were shown to be more like animals than plants in several key ways.

So, biology grew from centuries of having only two kingdoms – just two initial slices of the pie – to having three in the 19th century and four by the first quarter of the 20th century. Then, in the last quarter of the 20th century, that number went up another 50%. Systematics had evolved from being binary – just two kingdoms, plant or animal – to being sexinary (plant, animal, fungi, bacteria, archaea, and protists). In what follows, we'll learn more about how those new kingdoms were determined to be worthy of that title, and we'll see how systematists almost painted life into another binary corner.

A new systematics layer

Besides kingdom numbers, what also seemed to escalate in this era of change in systematics was the bickering among systematists. Each systematist had his or her own take on how alike or different various life forms might be. There was general agreement that two species would need to be very different from one another to be put in different kingdoms, but "very" is in the eye of the holder. Hence, some disagreements within the priesthood.

In the last quarter of the 20th century, a new method in biotechnology began to provide more objective and quantifiable data – DNA sequencing – also called genetic sequenc-

Plant or Animal?

ing, DNA profiling, and DNA fingerprinting. In DNA sequencing, selected genes or entire gene sets from different species can, in essence, be laid side by side and compared for their degree of similarity. The more similar the DNAs are, the more closely related the species are assumed to be. The more unalike they are, the farther back in evolutionary time (and the farther down on the Tree of Life) one would need to go to find a common ancestor.

Systematists who used this new approach began to see that important differences in life forms were evident above the kingdom level – distinctions that were not being captured in the Linnaean system. So a new, highest-level taxonomic category was run up the flagpole. Many systematists now salute that new flag. Its name is domain, or sometimes dominion or occasionally super kingdom, empire, or realm. While most systematists appear to be okay with the concept of a nomenclatural rank higher than kingdom, some insist on using their own nomenclature for it.

We could envision a couple of domains in our mode-of-transportation classification scheme if we added the approximately 7.7 billion sometime pedestrians plus those who move around on animals' backs. Pedestrians and camel riders are so different from the machines associated with land, water, and air travel that we might want to put them into a domain of their own – maybe the Peds. But, then we would need a domain name for the three non-ped kingdoms of transportation. They could be Mechanicals.

It was perhaps easy, once genetic fingerprints and other data were in front of them, for systematists to identify three domains: Bacteria, Archaea, and Eukarya. That nomenclature was proposed by Carl Woese and others in 1990. Bacteria will sound familiar. They have been around for close

The Gyroscope of Life

to 4 billion years. They started out as what I called Life 1.1. Archaea, which make up a second domain, emerged from bacteria as Life 1.2. They look like bacteria, but they differ in some key biochemical and genetic traits. Those differences are so great that many systematists feel bacteria and archaea belong in separate domains. After all, the two have been going their own way evolutionarily for 2 or 3 billion years.

Since 1925, the monales, which included both bacteria and archaea at that point, have been described as prokaryotes. The prokaryote name was suggested by the same Frenchman, Edouard Chatton, who rebranded the monales as a kingdom in 1925. He suggested all other life forms, besides monales, be called eukaryotes. The distinction, which was easily determined as microscopes became more powerful, focused on how cells package their DNA – their genetic material. In prokaryotes, DNA just sloshes around inside each cell. In eukaryotes, DNA is organized into units we call chromosomes, and those chromosomes are all packaged into a membrane-bound nucleus. That is a huge difference, of course. It undoubtedly reflects evolutionary changes that go back at least a couple billion years. That is ample argument for putting prokaryotes and eukaryotes into different domains, and they were as time and systematists wore on.

Systematists could potentially have created just two domains – Prokaryota and Eukaryota. A binarity. That would have given recognition to the fact that life comes in two distinct forms at the cellular level – nucleate and anucleate. In fact, Carl Woese, who proposed a three-domain system in 1990, first described the prokaryotes as a domain in a 1977 paper. (Woese, a microbiologist, established the archaea as their own kingdom in 1977.)

I've only half-paid attention over the years to what the

Plant or Animal?

high priests are saying. Much of what they say tends to be reversed or superseded rather quickly. But, if I follow the arguments properly, Woese and his colleagues made the right three-domain call in 1990 and for solid scientific reasons – not just to sidestep another binarity. From all indications, bacteria and archaea are almost as different from each other as we eukaryotes are different from bacteria. Indeed, the archaea are more like us in several genetic and biochemical ways than they are like the bacteria, causing systematists to suggest that we eukaryotes evolved from the archaea side of the Tree of Life.

So, it seems reasonable to describe life as occurring in three distinct domains. Some systematists, by the way, are making noises – even pronouncements – about more domains. Some would put viruses into the Tree of Life as another domain. Some have even suggested grafting prions and viroids on as a fifth domain. Stay tuned...

Using the new genetic technology, systematists also confirmed that the three well-established eukaryotic kingdoms (animals, plants, and fungi) are distinct from one another. Each kingdom has unique hardware and software features that reflect their separate evolutionary paths. Genetically, Protista and another couple would-be kingdoms turn out to be a mishmash of eukaryotic forms, probably just evolutionary jetsam and flotsam. That has been the kiss of death as far as many modern systematists are concerned.

Recalculating

As noted earlier, a pretty good case has been made for projecting the existence of 8.7 million species of higher life forms. But that is just the tip of a huge biological iceberg.

The Gyroscope of Life

We "higher life forms", the eukaryotes (plants, animals, and fungi), make up an embarrassingly small number of species when we bring into focus microscopic life forms. Microbes, the bacteria and archaea, have been evolving for much, much longer than eukaryotes, and they can be found in just about every nook and cranny on Earth, to include nooks and crannies on and in perhaps every plant and animal. A bit of autobiologic, then, suggests microbes should be more abundant species-wise. We already know they are hugely abundant just in terms of sheer numbers of individuals. Each of us is made up of around 3 trillion cells, and we each walk around with about 3.8 trillion microbes in and on us – hosting perhaps over 1,000 "microflora" species, mostly good/beneficial and mostly in our guts.

Using new DNA-sequencing technologies to detect the presence of previously unknown microbial species, one estimate (from biologists at the University of Arizona in 2017) raised the likely number of current species to 2 billion, with microbes constituting around 80% of that number. A 2016 report from Indiana University estimated that microbial species alone might number from 100 billion (or 0.1 trillion) to 1 trillion. That mind-blowing estimate, which some now consider conservative, is also based on DNA-sequencing results. Analyzing samples from soils, oceans, aquifers, insect guts, showerheads, etc. reveals millions of microbes so distinct from one another that they must, as a minimum, be different species. Many of them appear so different that they have been tentatively placed into new phyla. We know next to nothing about these never-before-reported microbes except that they are there. Microbiologists are working hard to discover, describe, and name as many of them as they can. Some of the urgency comes

Plant or Animal?

from a profit motive. The hope is that some of those 0.1 or 1 (does it really matter?) trillion species will have biochemical tricks up their sleeves; tricks that will have medical uses, help solve environmental problems, or make them bio-factories. Talk about a huge task. Talk about huge potential pay offs – both societally and entrepreneurially. So finding and naming living things has value.

The perhaps-not-too-hypothetical figure of 0.1 to 1 trillion living species comes almost entirely from estimates of the number of bacteria and archaea. Using that figure and the 8.7 million estimate for the number of higher life forms, we find that the eukaryotes make up about 0.00087 to 0.0087% of all species. We are a very, very thin sliver of the mincemeat pie – a tiny twig on the Tree of Life. But that's a highly significant 0.00087 to 0.0087%. It includes all macrobiota – the plants, animals, and fungi that make up all of visible life.

We will return to these themes in a chapter that discusses species and evolution. For now, the key point is that it took many centuries to establish that living things do not constitute a binarity – not just chocolate or vanilla, not just plant or animal.

Heterotroph versus autotroph: Another binarity?

With apologies, I need to introduce two more technical terms describing living things: heterotroph and autotroph. These words are quite useful and quite impossible to render in a few more-familiar words. Heterotroph and autotroph aren't taxonomic layers, like domain or kingdom, but they are used to describe some important characteristics of living

The Gyroscope of Life

things. In our belabored modes-of-transportation classification scheme, these terms would be somewhat analogous to noting which cars, boats, or planes are powered by fossil fuels and which use renewable-energy sources. For example, a car or a plane can be powered by petroleum products or by photovoltaics, i.e., the sun. (In 2016, a strictly solar-electric plane completed a circumnavigation of the Earth.) Likewise, a ship can be powered by fossil fuel or by wind. Putting a name to its energy source doesn't cause us to place a car, plane, or boat in a different hypothetical taxonomic category. It just provides us a useful piece of information about how various transporters are being powered.

Among living things, most organisms take in complex chemicals and break them down to get the energy and building materials needed to carry on life. Such organisms, which include us, are called heterotrophs (Greek for other and feeding). We are heterotrophic (adjective) and live by heterotrophy (another noun to describe this energy-obtaining life style). Heterotrophs are found in all three domains: Bacteria, Archaea, and Eukarya. Among the eukaryotes, all animals and fungi are heterotrophs. The great majority of bacteria and archaea are heterotrophic too. We heterotrophs eat (or take in in some fashion) energy-rich, complex chemicals (food) and use both the chemicals and energy stripped out of them to grow, maintain, and spin our cellular and organismal Gyroscopes of Life.

An autotroph (from the Greek for self and feeding), on the other hand, nourishes itself by obtaining matter and energy in simpler forms and using them to grow, maintain and spin its cellular and organismal Gyroscopes. Most autotrophs are photosynthetic and are powered by sunlight. Green plants capture energy from sunlight, take in simple chemi-

Plant or Animal?

cals from air, soil, and water, and then make complex organic (carbon-containing) chemicals. Most plants are green and autotrophic, although there are some interesting exceptions to be discussed shortly. A small percentage of bacteria and some protists also have photosynthetic capabilities and likewise rely on sunlight to provide energy to power their lives.

So, a cow (a heterotroph) eats leaves of grasses (autotrophs) and breaks the leaves' chemicals apart to extract their energy. The cow can then use that energy to keep herself warm, to repair stuff that happens because of the 2nd Law, and to make other complex stuff – milk, filet mignon, and calves. We (also heterotrophs) can drink that milk or eat that steak and get chemical energy to warm and repair ourselves, to grow, and perchance to make babies. If we connect all those dots from leaves of grass to babies, we can see that all – from green grass to cooing babies – are using energy ultimately from the Sun to stay alive. Food chains typically have plants at the bottom, plant eaters next, eaters of plant eaters next, and then eaters of eaters of plant eaters. The plants are autotrophs. Everything above is a heterotroph. A trophic binarity.

One of the interesting exceptions to food chains with photosynthetic autotrophs at the bottom is seen around deep-ocean hydrothermal vents. These peculiar geysers of super-heated, mineral-laden water occur 2 or 3 miles below the ocean's surface and are home to some unique creatures. No light makes it down there; we need not look for photosynthetic forms to be at the base of the food chains. However, some types of archaea living around the vents can use simple minerals coming out of the vents to get the chemical energy they need to make more complex stuff – and more of themselves. These chemoautotrophs constitute the first level

The Gyroscope of Life

in deep-ocean-vent food chains. They are eaten by heterotrophs, which are eaten in turn by other heterotrophs, and so on… Still a trophic binarity.

Which came first: Autotrophs or heterotrophs?

Let me pose a riddle. All food chains that ecologists have examined begin with autotrophs at the bottom and then heterotrophs at all successive levels. Today's heterotrophs are clearly dependent on autotrophs for their existence. In many creation accounts and in both biblical creation accounts, plants were created before animals. In the straightforward day-by-day recounting in Genesis 1:1-2:3, plants were put on the Earth 3 days ahead of animals. In the more poetic Garden of Eden story in Genesis 2:4-25, Adam was formed and put in the Garden that had been planted for him. Then all of the animals were formed in a search for Adamic companions. Each of these accounts offers what would seem to be the logical order of appearance: autotrophs and then heterotrophs.

But the chemical fossil record doesn't necessarily support an autotrophs-first sequence. Chemical fossils provide evidence of life going back to about 4 billion years ago (bya). All those life forms appear to have been heterotrophic – or at least not photosynthetic. The first evidence of photosynthesis appears in rocks about 3.5 billion years old. So, before there were photosynthesizing autotrophs, how could life have proceeded for 500 million years without an energy source for primordial food chains? Trick question. The answer is: it couldn't. The 2nd Law would have caught up with such a system and shut it down quickly. There must be a continual supply of energy to maintain the order necessary for life – to

Plant or Animal?

keep entropy low and to keep Gyroscopes spinning.

So, what was the energy source for those first life forms 4 bya? The answer is – no surprise – nobody knows. A couple of different hypotheses suggest themselves. One is that the primeval environment had lots of moderately complex organic chemicals floating around in its oceans, and the first life forms were able to mine those chemicals for energy and as building materials. The primordial soup likely had amino acids, simple sugars, nucleotides (building blocks for DNA and RNA), and other energy-rich molecules. Those chemicals, usually thought of as biochemicals, could have been formed by abiotic processes both on Earth and in space. In labs, chemists have recreated conditions that are thought to have characterized early Earth. They used lightning-mimicking electrical sparks and geothermal-like heat to spur on chemical reactions. Those conditions produced more-complex, carbon-containing chemicals. (The 2nd Law was not violated; electrical and heat energy were being supplied to create chemical energy and generate order.) Similar chemistry is thought to occur in interstellar space. The products can be some fairly complex organic chemicals that make their way to Earth in meteorites and comets. More star dust. So, one theory for how heterotrophic life might have been able to succeed and proceed in the absence of autotrophic life is that there was a soupy reservoir of complex chemicals for the heterotrophs to "eat" and live on.

A second theory, or hypothesis, to explain a heterotrophs-first fossil record is that some of those first forms weren't actually heterotrophs. They weren't photosynthetic autotrophs – not until 500 million years into life. But they may have been archaea-like beings that were chemoautotrophs – able to strip energy out of simple minerals. We know that archaea

The Gyroscope of Life

are at the base of food chains and food webs in deep-ocean hydrothermal vents. If life originated down there, and some suggest maybe it did, we would not be faced with the conundrum of how could heterotrophs be first? Maybe they weren't first. Maybe that title goes to chemoautotrophs.

If you were a fan of "Big Bang Theory", a popular television series, and, if you paid attention to the series' theme song, you may have detected something quite related to this discussion. The catchy theme song, composed and sung by the Barenaked Ladies (an all-male group), has a couple of lines that go: "The Earth began to cool. The autotrophs began to drool." I was willing to grant the Ladies some poetic license, but I couldn't think of any drooling autotrophs. Then I remembered the sundews, members of the genus *Droscera*, which includes almost 200 species of insectivorous plants. They are called sundews because they produce a clear, perhaps drool-like, secretion on the tips of hairs on their leaves. The secretion sparkles in sunlight, giving the plants their common and scientific names. *Droscera* comes from the Greek for dew or dewdrop. That "drool" can attract and trap insects. The sticky hairs can then curl to bring snared insects into closer contact with the leaf. Then the insects get digested by juices – more drool – secreted by the leaf. Sundews carry on photosynthesis. They are true autotrophs, with a drooling problem, and something of a heterotrophic tic. So, Ladies, sing on. But we are still looking down the barrel of another biological binarity, something that Mona seems to dislike.

Crossovers and binary busters

Some life forms appear on first blush to be both hetero-

Plant or Animal?

trophs and autotrophs – binary busters. These are living things that have both autotrophic and heterotrophic features, even more so than sundews. Lichens are interesting crossover life forms that can live in some unpromising places, such as on bare rock. Lichens were once considered distinct organisms, members of particular species. We now know lichens are marriages between two species, an autotrophic alga and a heterotrophic fungus. It is a marriage of convenience, with the fungal partner being the primary beneficiary. The fungus maintains tight control over the purse strings in these marriages. As much as 80% of the sugars produced by the alga ends up in the fungus.

Corals are in the animal kingdom, and therefore heterotrophs. But they have a history of being a little confusing. Aristotle called corals "zoophyta" (animal-plants), because of their simultaneously animal- and plant-like characteristics. They are still commonly referred to as zoophytes. Systematists seem to have waffled a bit when they put corals into a group called Anthozoa, which comes from the Greek for flower and animal. But, in fact, that name reflects the beautiful flower-like colors and shapes of some corals. So-called hard corals produce limy coverings or skeletons around their living tissues. Over many decades or centuries, that process gives rise to coral reefs, which are so important in marine biology.

Many corals are strictly heterotrophs, but some species of coral have developed an association with algae that is reminiscent of what goes on in lichens. Under the right conditions, algae can take up residence inside corals and carry on photosynthesis. Up to 30% of the living tissue in some corals can be algae. It is not clear how mutually beneficial this arrangement is. The algae appear to benefit from the carbon dioxide and nitrogenous wastes produced by the coral. The coral may obtain some sugars or other complex forms from the algae. What we know, though, is that coral under stress will eject the algae. There must be some cost to the coral for carrying on the relationship, at least when under stress. When the algae are ejected from the corals in a reef, the whiteness of the limy skeletons becomes obvious. The algae-ejecting phenomenon is known as coral bleaching. It is considered a sign of coral stress. The bleaching can reverse if conditions improve and algae reenter their coral cabanas. If the stress continues, the coral can die. This is a major concern for marine biologists and ecologists. The major stressors that seem to be causing coral bleaching are increasing carbon dioxide levels, which make seawater more acidic, and waters warming as a result of global climate change.

With both lichens and algae-harboring corals, two organisms – one autotrophic and the other heterotrophic – are cohabiting symbiotically, but each can live alone as it turns out. Symbiosis comes, of course, from the Greek and means living together. Except for some insectivorous plants, which are marginal exceptions at best, autotrophy and heterotrophy seem pretty binary. But more is coming…

Plants parasitizing plants: Botanical vampires

Here are the real autotroph-heterotroph binary busters: some plants have become partial or total parasites – heterotrophs – over evolutionary time. Today's land plants have had about 450 million years of natural selection's trial and error to come up with ways to obey Mona's House Rules, and some of them have stumbled onto parasitism. By some estimates, maybe 1% of flowering plants are parasitic to

Plant or Animal?

some degree. These are species whose ancestors once lived on their own as fully autotrophic, self-supporting, seed-bearing plants. But somewhere along the evolutionary path, these plants wandered toward the dark side and developed an ability to attach themselves to another plant and extract its resources: water, nutrients, sugars, and other complex biochemicals. That heterotrophic tack has served them well, but it often messes seriously with their hosts.

Some plant species are take-it-or-leave-it parasites, also being able to live fully on their own as straightforward autotrophs. Among my favorite wild flowers are the Indian paintbrushes, also called prairie-fire. The paintbrush genus has about 200 species, all of which are able to live by autotrophy or by sponging some of their needs off other plants. They are natives of the New World with short stems and spikes of red, pink, orange, purple, white, blue, and yellow (and just about every shade in between) flowers. Most have green stems and ample green leaves – enough to make it on their own as full-fledged autotrophs. But they often attach their roots to the roots of grasses or forbs and draw some portion of their nutrition from those hosts. They are generally not considered to be particularly harmful to their hosts – more like a nice, over-staying house guest than someone who eats you out of house and home.

The mistletoe that we know for its party-improving qualities is a partial parasite. Its green leaves tell us that it is not wholly parasitic. It is not considered harmful, primarily drawing on its host's water supply. But some of its relatives – also often called mistletoes – have gone totally to the dark side as parasites, growing on host plants and producing no green tissues. They can be injurious, even fatal, to their hosts. Parasitism is apparently such an easy shortcut to completing

The Gyroscope of Life

one's life cycle and passing on one's DNA that other plants have also veered entirely to the dark side and stopped making green leaves and photosynthesizing. They have become able to live only by parasitism, making them fully heterotrophic – relying entirely on complex materials they pilfer from their hosts/victims. Some of these vampire-like plants have become major pests and make the worst-ever house guest look like a nice guy.

Broomrapes are some of the most devastating parasitic plants. The broomrape genus has more than 200 species. Most are pretty, bearing clusters of blue, white, or yellow flowers on a single stem. The stem is generally pale yellow and produces no leaves, and that is a red flag. Broomrapes don't need green parts, because they are 100% parasitic. A broomrape seedling attaches itself to the root of a host plant and burrows into the plumbing that transports water, minerals, and complex biochemicals. It then can rely totally on the host for everything needed to make a stem, pretty flowers, and seeds. Those seeds may then lie dormant in the ground for many years until the root of a suitable host grows near enough for the broomrape seedling to latch on. The cost to the host plants of supporting the broomrape vampires can be so great that, in severe agricultural infestations, an entire crop can be lost. Crops subject to attack by various broomrape species include potato, tomato, pepper, cabbage, sunflower, and bean.

Witchweeds are in the same family as the broomrapes but belong to a different genus. Compared to the broomrapes, witchweeds are equally pretty and equally or more devastating. The broomrapes are from the northern hemisphere, while the three dozen or so witchweeds are native to Africa, southern Asia, and Australia, but each genus has crossed into

Plant or Animal?

the other's territory as a result of agriculture. Witchweed is also totally parasitic. The witchweeds have a strong affinity for members of the grass family. The grasses affected include corn, sorghum, wheat, rice, sugarcane, and more. Especially for subsistence farmers, a witchweed infestation can be devastating. This year's crop is lost, and the seeds produced will infest the ground for 10 years or more. Some sophisticated – read expensive – mechanical and chemical methods of witchweed control are available, but subsistence farmers have little choice but to seek out new, uninfested sites.

One of the most phantasmagorical plant parasites is corpse flower – also called corpse lily – that produces the world's largest flower. The couple dozen species of corpse flower all come from Southeast Asia and Oceania, and all are total parasites – heterotrophs – with no green tissues. The most outrageous species of these plants is one that parasitizes tropical vines related to grapes. The parasitic plant produces no visible aboveground tissues at all except a single, humongous, fleshy, smelly flower – often over three feet wide and weighing up to 24 pounds! That's one hell of a flower. The flower bud pushes its way out of the ground and then opens to carry out its reproductive role. The flower's powerful rotting-flesh smell earned it the name of corpse flower. That smell attracts pollinators and creatures that carry seeds to new locations. These parasites are not a big pest where they occur, but they typically create a big news story when one flowers in a greenhouse the public can visit. People will flock to see the huge flower – even while holding their noses.

If you get to visit such a greenhouse, just know that there are two different plants that have the common name of corpse plant or corpse flower, and each of them is sometimes billed as the world's largest flower. The other one, also

The Gyroscope of Life

known as the titan arum, is an inflorescence (a cluster of flowers), and it is a green plant – not a parasite – in the same family with calla lily and jack-in-the-pulpit. Titan arum is quite impressive in its own right, however, producing a smelly flowering structure that can be 10 feet tall and leaves that can reach 20 feet long.

Mycorrhizae and plants parasitizing fungi

One of the more intriguing and satisfying stories about parasites is one where the worm turns, and a parasite becomes the one parasitized – by its former victim. This story must begin with a bit of background about another Greek-derived term: mycorrhizae. Translating the Greek gives us fungus-root. The word refers to a widespread symbiotic phenomenon. Better than 90% of plant families enter into mycorrhizal relationships, where fungi and plant roots form a super-tight bond. In many cases, the fungal partner completely ensheathes portions of the plant's roots with a mass of filaments. In other cases, the fungal filaments invade the roots in a pattern that is reminiscent of how fungal pathogens attack roots. In fact, some mycologists speculate that today's beneficial-to-both-parties mycorrhizal relationships started out a few hundred million years ago as a pathological relationship, where the fungi were parasitizing the plants. Over time, they reached a mutually beneficial détente.

Mycorrhizae are a win-win symbiotic collaboration between a plant and a fungus. The fungus provides key services to the plant, to include enhanced water and nutrient absorption. For these services, the fungus receives due compensation in the form of sugars and other complex biochemicals. Growing plants with or without a fungal partner produces

Plant or Animal?

striking differences. The mycorrhizal plants are always larger and healthier. In many cases, the symbiotic plants are two or three times as large within a year or two. The fungus-plant association, which may have started out as a disease, now provides great advantage.

What we have thus far established about mycorrhizae is that the plant-fungus symbiosis, which may have begun as a pathogenic relationship, provides a great leg-up to plants today. In fact, the boost is so great that most higher plants take advantage of it. The orchid family is particularly noted for using the mycorrhizal edge. Orchids have evolved not only to enjoy the mycorrhizal partnership, but to totally depend on the fungus at key times in their life cycle.

Orchids are the first or second largest family of flowering plants, having more than 25,000 species worldwide. They are notable for their beautifully sculpted flowers and amazing mimicry of insects to lure in pollinators and do their birds-and-bees thing. We'll have much to say about orchid sex in the next chapter. Orchids often have large, sexy flowers. But the seeds of orchids are minute – little more than specks of dust. They are produced in prodigious numbers. The seed capsules of some orchids may make a million of these motes. That size and those numbers are immediate hints that orchid seedlings might have a hard time taking hold. They do. Sometimes none of those million seeds will result in a single new plant. The difficulty they face has to do both with their size and with their dependence on help from mycorrhizae to get a jump start.

An orchid seed is a rudimentary orchid embryo with no energy-rich materials on board. That means that the little seedling will have no reserves to get itself established as an autotroph. Seeds of most plants – but not orchids – carry on

The Gyroscope of Life

board a supply of energy-rich materials that help the seedling produce its first roots, stems, and leaves. Once those leaves are above ground and photosynthesizing, the seedling can become self-sufficient. Until then, the seedling is dependent on the energy supply in the seed, or on energy that can be scrounged from another living thing. All orchids have come to rely entirely on fungi to provide them with the energy needed to get their seedlings established. The helpful fungi get their energy from decomposing organic matter in the soil or on tree bark. Many orchid seeds will not even begin to germinate unless they come in contact with a fungus that they can parasitize to receive a mycorrhizal boost. This worm-has-turned relationship is sometimes described as mycorrhizal cheating. The young orchid plant is a parasite on the fungus. It is a heterotroph.

For most orchids, the period of parasitic heterotrophy lasts only until the orchid seedling can produce its own energy supply via photosynthesis – becoming an autotroph. Once the orchid plant has developed sufficient leaves thanks to the jumpstart from the fungus, the roles of the partners can be realigned. However, it is not entirely clear with some orchids just how distinct those roles become. In some cases, it appears the orchid continues to rely on the fungus for some of the complex organic materials that it (the plant) needs – along with water and nutrients. The fungal partner may never get into a relationship where the plant provides it with sugars and other complex materials. That was the original mycorrhizal arrangement, but the rules seem to have changed with the orchid cheaters.

In the early 1970s, in my military years, my wife and I were traveling through Switzerland and stopped to do some botanizing in the forests on the southwestern side of Lake

Plant or Animal?

Thun. These were rich, diverse woods with many hardwoods and an interesting understory as well. The most interesting thing we found in the understory was a strange plant that was about 10 inches tall. It was just a single, pale stem with no leaves and several pale flowers – no distinctive coloration to any of the parts. The flowers weren't large or showy – especially since they were almost translucent. They were clearly orchid flowers, because I had seen many of their peculiarly shaped cousins over my years of botanizing. But what was totally different about this orchid was that nothing was green. This plant reminded me of ghost plant, or Indian pipe, a pale white parasitic plant in Appalachia. That got me to thinking something similar might be going on with the plant I was looking at and photographing there in Switzerland.

When I got back to my books on European flora, I flipped to the orchid family and pretty quickly found that I had been looking at a bird's nest orchid. I learned that it is common and that it can have some brownish coloration. The one I saw was growing in some pretty deep woods and perhaps was a little paler as a result. What I have learned since is that bird's nest orchid is one of several orchids that have totally abandoned autotrophy to become life-long parasites (heterotrophs) on their former mycorrhizal ally. They live entirely off of fungi, which are in turn living off of dead leaves and other detritus. Interestingly then, these plants are at the third level of this food chain. Dead plant matter (so autotrophs are at the bottom) is "eaten" by fungi, which are the second level. Then these parasitic plants live off of the fungi.

What's the take home? Autotrophy and heterotrophy are not binary. Most plants are totally autotrophs, but some plants are totally parasitic heterotrophs, and some plants

The Gyroscope of Life

have learned to straddle the line. Some straddle the line by being partially heterotrophic throughout their life cycle (paintbrushes and mistletoe). Others – especially orchids – spend part of their life as heterotrophs and then come back from the dark side to be self-sustaining autotrophs.

8

Male or Female?

Urge, and urge, and urge;
Always the procreant urge of the world.
Out of the dimness, opposite equals advance —
Always substance and increase, always sex.

Walt Whitman (1819-1892)

Why I might be particularly well-qualified to talk about sex and gender

You will learn why shortly, but know for now that I am castrate. There's no seamy or court-driven reason. It was a medical choice, and it happened while I was writing this book. In fact, I was recovering from "that little procedure" as I revised this. Suffice it to say that having no nuts gives me a somewhat different perspective about sex and gender. But, rather than disqualifying me from talking about such matters, I think it helps me to see them in a clearer, less biased way. I still consider myself a guy, but testosterone poisoning is not at play anymore. It hasn't been for over eight years, also for reasons to be explained shortly. You could consider me a third party in a discussion of gender. But, in fact, I've looked at sex and gender from a biological perspective for over 50 years. That perspective perhaps takes a little glimmer off of the subject, but it sheds some light at the same time. Demystifying sex and gender is not automatically a good thing, but it can be if the mysteries have led to misunderstandings.

The Gyroscope of Life

Gender and sex in the animal kingdom

Mona's House Rule #3 is "pass on your DNA, make love and make babies". I called it the reproductive imperative. It's more like the Prime Directive. Most higher-life forms (animals, plants and fungi) go to considerable lengths to make babies, and all must operate within a two-gender system to do it. Or must they? This chapter is going to look more closely at the concepts of maleness, femaleness, and sexual reproduction. As it turns out, male or female is another blurry binary, just like dead or alive. Gender as a binary seems right. It's the biblical way: Genesis 1:27 explains, "God created man in His own image, in the image of God He created him; male and female He created them". That is the fundamental theology of gender. But the biology of gender is not always so clear cut – not for us or for other creatures.

Freemartins are a common example of blurred gender. Freemartins are cows that exhibit bull-like features and behaviors. Masculinized cows were described in Roman times. Not until the 17th century, though, did someone notice – or at least document – that freemartins always have a male twin. For most of the time since then, it was supposed that the freemartins' masculine characteristics were due to crossover hormonal influences during the shared time in their mom's uterus. In the early 20th century, however, it was shown that these female cattle have some male cells, which cross over via interconnected placentas from their bull twin, and those male cells override many female features as the cow develops. Freemartins are the norm for a twin cow with a male womb mate. Freemartin effects are sometimes also seen in sheep, goats, and pigs. No cases in humans have been

Male or Female?

reported, literary references notwithstanding, and there is no known flip-side effect – where a male twin develops feminized features.

When we look at it even casually, autobiologic suggests that gender determination and sexual development – making a little boy or a little girl (or a bull or a cow) from a fertilized egg – can't be simple. Many, many steps and billions of cells are involved. At every step along the way comes the potential for something to go wrong, and we know what Murphy's Law suggests about that. It would seem almost inevitable that maleness and femaleness might not be absolutes, and they aren't. Freemartins are just one example.

Most are familiar with the notion that human gender is determined by the pairing of X and Y chromosomes – the sex chromosomes. Girls have two X chromosomes, one from Mom's egg and one from Dad's sperm. Boys get an X from Mom's egg and a Y from Dad's sperm. But, as Murphy predicts, things can go wrong even at this first step of making little boys and little girls. Chromosomal mix-ups can occur when eggs and sperm are being made and when a newly fertilized egg begins to divide. One of those mix-ups can result in an extra X (female) chromosome being present – XXY. The offspring are male. Their only symptom may be sterility. The XXY condition may in fact go undiscovered, but it is thought to occur in one or two births per 1,000 boys born.

The XYY chromosome combination is a one-in-1,000 occurrence with an interesting history. These boys have an extra Y (male) chromosome. Due to unwarranted speculation by the 1960s discoverers of the condition and compounded by even worse reportage, XYY males were mislabeled as "super males". They were expected to be twice as ornery as ordinary guys. After all, they carried two male chromosomes. Dur-

The Gyroscope of Life

ing the early 1970s, such boys and men were thought likely to be more aggressive and violent, and those discovered to have the XYY condition underwent special counseling. Some criminals used the "XYY defense" in efforts to absolve themselves of violent acts. Fortunately, juries and jurisprudence did not buy that defense. Good science has been done since then, and it shows no correlation between being XYY and being violent. Men who are XYY tend to be about three inches taller than average, and they may have problems with acne, but they are otherwise, on average, quite average.

Turner syndrome, which occurs when there is a single functional X chromosome, can be devastating. It can be due to mix-ups either in egg or sperm formation or in the first few divisions of the fertilized egg – Murphy's Law stuff. With no Y chromosome and therefore no genes for maleness, the individual will be female. Girls with Turner syndrome may appear normal, but multiple serious medical problems can occur. Those can include atypical sexual development and heart problems. Turner syndrome is present in about one in 2,500 live births, but the great majority of conceptions that produce Turner syndrome fetuses will end in miscarriage – all because of a miscue in the first step of gender determination.

Other less frequently observed sex chromosome combinations include XXX, XXXX, XXXY, and XXYY. Some produce relatively minor effects, but others can be life-altering and life-shortening. All of this is to say that the gender lottery doesn't have just a binary – XX or XY – outcome. Or, to use another gambling metaphor, the gender roulette wheel has mostly male and female pockets, but the ball can occasionally land on something else. Or, if you don't gamble, just know that the chances of anyone being a straight-up,

Male or Female?

no-questions-asked, no-doubt, strictly male or female are not 100%. Binarity can be "violated" at the very first step in determining gender.

But gender determination is much more complicated than just X and Y chromosomes, even when there is a clear-cut outcome in the gender lottery. Much has to happen for an XY zygote (the egg after it has been fertilized) to produce a baby with testes and a penis. Likewise, it's a long way from an XX zygote to having a baby with ovaries and a vagina. It's not like the sperm delivers a perfectly formed, miniaturized version of a baby boy or baby girl. That was an accepted notion in the pre-scientific 16th century. According to that thinking, sperm contained a homunculus – a little man (or woman) – that grew in the inseminated woman to produce whatever the gender of the homunculus was. The whole notion of a woman being seeded with something that grows into a child still clings to our lexicon. Semen comes from the Latin – not Greek! – for seed.

For a while in the 20th century, femaleness was thought to be the default or "off" condition for an embryo or fetus. A male "on" switch on the Y chromosome presumably triggered maleness. Without that switch being flipped, the embryo developed into a female fetus. As of the 21st century, it looks like maleness must be switched on and femaleness must be switched off in order for a male fetus to develop. That switching is genetically and hormonally controlled in an intricate series of steps and events in the womb and on into puberty. If all goes just right, we get a masculine male from an XY zygote. And, if all goes equally well at all of the critical steps, we get a feminine female from an XX zygote. But mutations in the genes controlling many critical sex-determining steps and Murphy's Law screw-ups in the

The Gyroscope of Life

biochemistry of those steps predictably cause anomalies in the formation of sexual organs.

About one human infant in every 2,000 live births is intersex, or hermaphroditic, with reproductive organs that are not distinctly male or female. (In one report, that number was recently reported to be closer to one in 1,000 births.) In the US, that more conservative rate amounts to about 2,000 intersex births each year. Globally, the figure is over 65,000 annually. Over the centuries, these individuals have been celebrated in some cultures, but most modern societies go to great lengths to make intersex individuals either more fully male-like or more fully female-like.

Since the 1950s, ambiguity-removing surgery and hormonal interventions have been increasingly used on intersex children. Surgeons make intersex individuals look more like a little boy or a little girl. That is perhaps a logical and loving thing to do if gender is, indeed, binary. But the medical community has reached no consensus on how the imposed gender should be chosen. Should the decision be based solely on the child's chromosomal pattern or more on what Mother Nature has produced during its embryology? The choice has become rather politicized, especially when sex-change surgery is done during adulthood. We tend to deal with gender determination as a social or cultural issue – not as a biological phenomenon.

And then, there is the area of sexual identification and orientation. How do male brains get imprinted with sexy thoughts about females and vice versa? Is that hardwired? In the 1980s, it was common to hear that little boys and little girls are just alike neurologically. People who study brains and brain development, now point to differences in the structure and chemistry of little boys' and little girls' brains –

Male or Female?

as well as men's and women's. Those differences include gender-specific timing and distribution of gray and white matter during brain development. However, the brains of some boys and men shade toward female-associated differences, and the brains of some girls and women have male-associated properties. What are we to make of that? Is it Murphy's Law operating, or are social interactions causing changes to occur in boys' and girls' brains?

To muck up the discussion even more, environmental factors (for example, mother's medications during the pregnancy) can alter sexual development and sexual orientation, or identity. Diethylstilbestrol (DES), a synthetic estrogen, was used widely from 1938 to 1971 to reduce miscarriages and other complications in pregnancies. The FDA banned its use for such purposes in 1971, because it was shown not to have the desired benefits. Furthermore, it increased the risk of certain cancers for DES mothers and DES daughters. Up to one-third of DES daughters developed problems in their reproductive tracts, and several studies have documented that DES sons are more likely to develop gender identity issues and transsexualism. Autobiological conclusion: the data point to a likelihood that sexual orientation is not binary. Sexual identity is likely just as subject to variation and vagary as other developmental processes. Those who may have no gender identity problems can count themselves fortunate. Hopefully we can all be compassionate for those who have landed in a different pocket on the sex-and-gender-identity roulette wheel.

The pros and cons of castration

In the last few years, I have become more aware of the

The Gyroscope of Life

biological and psychological basis for the "procreant urge" Whitman described in "Leaves of Grass" (see opening quote). From early 2011 to just recently I have been chemically castrate as a treatment for advanced prostate cancer. I was given a shot every 6 months that shut down testosterone production in my nuts. The rationale is simple. Testosterone stimulates, among other things, the growth of prostate cells. When those prostate cells become cancerous and get into places they don't belong, testosterone still causes them to grow. To slow the growth of rogue prostate cells, testosterone production needs to be suppressed. Hence, chemical castration can prolong the life of someone with metastatic prostate cancer.

I never considered myself a particularly horny guy, but I was always aware of a certain attraction, tension, or urge – to use Whitman's word – that women could trigger. Let's just leave it at that. But, after chemical castration, that urge was gone. Poof! I had hot flashes, but they weren't nearly as interesting as the urge. The point is this: I can anecdotally attest to the power of a hormone, which is what testosterone is, designed to move along the process for making babies. It's not just about making sperm. A biologically driven psycho-social gender component is involved, and the driver's name for males is Testosterone.

But, as I suggested at the top of this chapter, I am not just chemically castrate. My testicles have been removed. It was an easy decision as far as I was concerned. There is no cure for advanced prostate cancer. I am never going to want testosterone in my system again. So, why not just cut those boys off? I could keep on getting the shots every 6 months, but they were beginning to hurt like hell for 3 or 4 days. Maybe my butt cheeks were developing sensitivity to

Male or Female?

the injection. The recovery from "that little procedure" my urologist performed was less painful than from those shots. And, when I did some cost comparisons, I found an orchiectomy – the medical euphemism (from the Greek, of course) for cutting off one's balls – was less expensive than just one of those damned shots. Yeah, Medicare is paying for it – and I really appreciate that! – but somebody has to pick up the tab. That somebody is the US government and its taxpayers. So, if you are a taxpayer, I've done my part to keep costs down. (You're welcome.)

Some other peculiarities and vagaries about sex and gender in the animal world

Let's consider some other ways of determining gender in animals before coming back to Xs and Ys in plants. In birds, some fish, some reptiles, and some insects, the sex chromosomes are labeled Z and W. They work much like the X and Y chromosomes, but with ZW being the female pairing and ZZ male. There is some indication in birds that the default body plan is male and that the W chromosome turns on development of ovaries, which then drive development of the female body type. In chickens, for example, if a hen's ovaries are removed or damaged, the bird will begin to develop male plumage.

An old Southern and Appalachian saying is, "A whistling woman and a crowing hen are bound to come to no good end." A friend told me that, when he was growing up, a crowing hen would be in the cook pot before the sun went down. There could be some biology behind the folk wisdom of shortening a crowing hen's life. If, in fact, the rooster-like behavior is triggered by ovarian problems, the hen's egg lay-

The Gyroscope of Life

ing days are perhaps over, and it might be time to have some chicken and dumplings.

It's not always about W, X, Y, and Z. Sometimes other factors determine gender or override the lottery's gender decision. The gender of some reptiles – and occasionally some birds – is determined by the temperatures at which their eggs incubate. Some fish can switch genders under changing environmental or social conditions. Clownfish – of which Nemo is the most famous – develop a social hierarchy with a female at the top. If she dies or disappears, a male fish can change gender and become the new egg-producing matriarch. (That sort of transition could make Nemo particularly hard to find.) Another fish found on coral reefs, the wrasse, does the opposite. A dominant male wrasse will collect a harem of females. If he dies or disappears, the largest female will change gender and take over as the head of the seraglio, to include producing sperm to fertilize the harem. The transition from egg-producing female to sperm-producing mail takes her-him about 10 days – some serious gender bending.

Some creatures, such as earthworms and many snails, are true flower children. They are flower-like at least – normally both male and female at the same time – hermaphrodites. They can co-copulate and inseminate at the same time they are being inseminated. But the banana slug, which is a large, shell-less snail of North America, is in a category all by itself. It can co-copulate with another slug or it can literally go fuck itself. What makes this self-copulatory act even weirder – and a bit painful to consider – is that sometimes the slug gets caught. I'm not talking about getting caught *in flagrante delicto*. I'm talking about getting stuck. The slug's penis can get trapped in the female aperture either of another slug or of a slug making love to itself. According to those who have

Male or Female?

observed banana slugs in this pickle, the solution is to bite off the offending penis. It's called apophallation. (Yes, there is a Greek-derived technical name for this phenomenon.) In cases where this problem develops between two slugs, the apophally can be a shared activity. In cases where the slug gets stuck on itself in more than just a figurative sense, it has to perform the apophallation. Banana slug mouths have rows of small teeth – hopefully sharp teeth – quite suitable for munching on mosses, dead leaves, and apparently slug penises. A banana slug penis does not regrow. The apophallated slug – okay, I made up that adjective – must spend the rest of its life in a strictly female role. Banana slugs can live up to seven years.

Some animals can make babies while getting around sex altogether. In this shortcut, babies are made from unfertilized eggs. This is asexual reproduction that produces clones (exact copies) of the mother. In XX female systems, all offspring will be female. In ZW female systems, the offspring produced asexually can be ZW female, ZZ male, or occasionally a WW female. The key takeaway, though, is that no hanky-panky is involved. The egg is triggered to develop into a baby whatever without needing sperm to get the ball rolling. It is a virgin birth of sorts. In fact, the name for this style of asexual reproduction – parthenogenesis (yep, from Greek) – means virgin genesis or virgin creation. Most examples of this way of doing business reproductively come from the lower echelons of the animal kingdom: some protists, some insects, some spiders, some snails, and some crayfish. But vertebrates – animals with backbones, and therefore considerably higher on the evolutionary ladder – have been observed to produce young this way too. Those include some sharks, some reptiles, some amphibians, some fish, and

The Gyroscope of Life

(rarely) some birds.

In those creatures for whom the gender determination apparatus has worked smoothly to produce discrete (but not necessarily discreet) males and females, making babies is biologically straightforward. It generally entails getting representatives from each gender to tango – even if, as in Nemo's family, someone has to change genders to get the two it takes to tango. If the tango is successful, a sperm will fertilize an egg. The zygote will become an embryo, then a fetus (in mammals), and then a baby – satisfying House Rule #3. Being testosterone free, that is about as sexy as I will get with the topic of animal sex. But, as we all know, sex can be more complicated (and more fun).

Sex and the single cell

Unicellular algae and other protists can hook up and reproduce sexually. But they don't seem to think about sex much. Most of the time, single-cell organisms simply divide asexually, producing two daughter cells. The new individuals are copies or clones of the original individual, and they can repeat the cloning process. That works well when conditions are good. Unicellular algae can quickly turn a pond green when the water is warm, the sun is shining, and plenty of nutrients are available. One makes two. Two make four. And so on.

But when conditions become iffy, some algae head to the bedroom. Maybe the pond begins to dry up, turn putrid, get too hot, or otherwise become unsuitable for floating around and making daughter cells. When they sense things may be going south, some single-celled organisms will look around for a partner, pair up, fuse, and comingle their DNA. That's

Male or Female?

about as sexy as sex gets for single cells. A resting-stage spore typically comes from this union – a spore that can resume life and return to a single-celled, asexual form when conditions again become favorable.

There is a logic to this way of only periodically doing sex. Asexual reproduction is easy. No drama. No dating sites to deal with. No awkward first kisses. And asexual reproduction makes copies pretty faithfully. However, over time, genetic errors (mutations) begin to creep into those cloned copies, and those mutated copies are then faithfully reproduced asexually. A load of genetic problems can accumulate until the clone is at a disadvantage – less able to fit a changed environment. (Mutations can occasionally be good, but most of the time they are liabilities. Mutations seldom improve genes that have been honed by evolution for a couple billion years.)

Sexual reproduction offers an opportunity to correct errors that may have crept in during the recess from sex. During and following sexual reproduction, partners' mutations can in effect be proofread against each other, and the fusion of two different gene sources can also produce new genetic combinations that make the offspring potentially more fit. In other words, sex provides survival value. With their set of genes repaired, it's safe for these unicellular forms to go back to the old clonal mode of reproduction once the pond gets livable again – at least for a few hundred more generations.

Seedy, but not flowery, sex

But, enough with the animals and single-celled creatures. Let's get into the sex lives of seed-bearing plants. We'll first consider briefly some important, non-flowering, seed-bear-

The Gyroscope of Life

ing plants: the conifers. Conifers include redwood, pine, fir, spruce, hemlock, cypress, and cedar. Conifers are characterized by producing naked seeds – not embedded in a fruit. The seeds are typically borne in a female cone, which is pollinated and fertilized by pollen from male cones. When pollen is being released by some pine species, the air develops a smoky goldenness with each puff of wind. It is a strategy that works well for pines: throw pollen grains into the wind by the billions. Some will find their way to a female cone. Fortunately, while that pollen can make our cars quite dusty, most of us (95% or more according to clinical studies) do not appear to be allergic to pine pollen.

The ginkgo, or maiden-hair tree, is an interesting relative of the more familiar conifers. Gingko's non-needle-like leaves look more like those of an oak or a maple. It is a native of China and has gained popularity in the US as an ornamental and medicinal. Ginkgo trees in China, where they have been grown as ornamentals and a food source for thousands of years, can be up to 160 feet tall and 2,500 years old. Ginkgo is interesting sexually. Each tree is either a pollen-producing male or an seed-producing female. Gender is determined just as it is for humans and other mammals – Xs and Ys – with the XY being male and the XX female. The gender of a seed is determined by whether the egg-fertilizing sperm (in a pollen grain) contained an X or a Y chromosome. The male ginkgo produces its pollen in cone-like structures. As with pines and many other wind-pollinated species, that pollen is released into the air in great numbers with good odds that some of it will find its way to a nearby female gingko tree.

All of that pollen from male ginkgo trees is quite allergenic to some people, so we might suppose it would be better not to have male trees in our yard. But female ginkgo trees

Male or Female?

can be problematic, too. The fleshy-coated seed that female trees produce and then drop by the hundreds can be messy and smelly. I've walked under a female ginkgo tree in the fall. The sidewalk was stained and slippery with the fleshy seeds, and there was a distinct odor of butyric acid. For those who don't know that chemical, think of vomit. That is the odor of butyric acid. So, those who want a beautiful maiden-hair tree gracing their lawn have to be willing to tolerate the allergens or the smelly mess. Most opt for the allergens. But, if there are no male gingko trees within hailing distance, a female tree can work to full advantage, because no gingko pollen is around to fertilize it and cause production of the smelly seeds. In the one report I've found on the subject, that sex-avoiding distance should be at least 0.1 mile.

Sex and gender in flowering plants

Flowering plants, the angiosperms, have active and often flamboyant sex lives. Flowers are all about sex and sexual reproduction, and sexual reproduction is all about making babies (as seeds initially) and moving the next generation onward evolutionarily. Gender plays a role in making babies for most – not all, as we will learn – flowering plants, but gender isn't as big a deal for plants as it is for animals. By design, nearly 95% of the angiosperms are both male and female. Most of them make flowers that have both egg- and pollen-producing parts. They are sometimes described as hermaphroditic, but there is no stigma or prejudice attached to the term in botany. In fact, two-sex flowers are described as "perfect", while flowers that are just male or female are "imperfect". We will discuss those perfect plants shortly, but first let's consider a few gender-specific plants, plants that, as

The Gyroscope of Life

imperfect as they are, are pretty much like us.

The genetic basis for determining gender is not known for most of the flowering species that produce separate male and female plants. One either-male-or-female plant that has been studied is white campion, an obscure but pretty little wildflower. It is one of about three dozen flowering plants known to be in the same XY boat with ginkgoes and mammals. Interestingly, perhaps more is known about the genetics of gender determination in campion than in humans. With the campion, three distinct genes on the Y chromosome – the male chromosome – are involved in determining maleness. Two promote maleness in different ways, and the third suppresses femaleness. In other words, genes on the Y chromosome cause the plant to go solely to maleness. Without a Y chromosome – in XX females – only femaleness can be expressed.

In addition to sharing an XY sex determination pattern with us, the white campion also must deal with what amounts to a sexually transmitted disease. A peculiar smut (a type of fungus) can infect the pollen-producing structures of the male campion flowers. Instead of producing pollen, the infected parts become dispensers of smut spores. Insects that come to collect nectar or pollen from infected plants carry away smut spores, which can infect the next campion plant the pollinator might visit. Flowers on female plants often become sterile as a result.

Two other X-Y-determined flowering species that might be more familiar are asparagus and marijuana. The XY-male asparagus plant is generally preferred in the garden, because it tends to be somewhat more productive than XX-female plants. Female plants invest significant amounts of their energy in making seeds and fruits – and not in making edible

Male or Female?

spears. With marijuana, just the opposite is true. The female plants are more valuable because of the higher content of THC-rich resins in unfertilized, sinsemilla (without seeds) female flower buds. In fact, the male plants are a liability. Marijuana growers typically remove male plants, which can make up half of a planting, as soon as they can be identified. That avoids the potential for pollen shed and the undesirable fertilizing of the female plants.

Other important or well-known flowering species with separate male and female plants – but not necessarily with X-Y determination – include some palms, some hollies, persimmon, pistachio, poplar, spinach, and willow. Methuselah, the Judean date palm grown from a 2,000-year-old seed, is a male. There are no female Judean date palms for him to fertilize, but he can fertilize female plants of closely related date palms, producing hybrids that are a mix of the two species. The plan is to cross Methuselah repeatedly with successive generations of the hybrids, introducing more and more of the otherwise extinct genes back into the later generations. After many iterative steps, these offspring can be 95% (or more) Judean date palm.

Another well-known – at least in internet circles – male-or-female species is bell pepper. One can find numerous reports that bell peppers (the fruits – not the plant itself) are either male or female. Male peppers have three "bumps" or lobes when viewed from the bottom; female peppers have four. Female peppers are sweeter and better for eating raw. Male peppers hold up better when heated and are preferred in cooked dishes. Or so you will find on the internet and discussed among some gardeners and foodies. It is all hokum. I do not doubt that some people find four-lobed peppers better tasting. I believe in the power of suggestion. But the

The Gyroscope of Life

biological facts are stacked against gendered peppers. Pepper plants are hermaphroditic. Both the plants and fruits are bisexual (in a somewhat different but mathematically correct usage of that term).

While about 95% of angiosperms produce female and male reproductive parts on the same plant, some of them go to considerable lengths to avoid self-pollination. (The reason a plant might want to avoid self-pollination takes about one lecture from Genetics 101, so please just accept that there are good reasons why a plant might want to avoid self-pollination.) One of the mechanisms to prevent a plant's pollen from fertilizing its own flowers uses onboard chemistry to keep stay-at-home pollen from germinating and growing. Pollen grains arriving from different plants of the same species are quite welcome, but same-plant pollen is not allowed to incestuously mix and mingle. The pollen either fails to germinate, or the pollen tube is prevented from growing, hence blocking any possibility of self-fertilization. Other hermaphroditic plants avoid self-pollination by separating pollen-producing parts from pollen-receptive parts. Squash plants have separate male and female flowers on the same plant. Some plants have flowers that put pollen-receptive surfaces out on the end of a stalk, while the pollen-producing parts are held close to the vest. Such strategies reduce the potential for pollinators to transfer pollen within the same plant or flower.

Avocados have an interesting way of avoiding self-pollination. An avocado flower opens initially as a female. It will remain in business as a female flower for a half day or so, and then it will close up shop. The next day, the same flower reopens, but under new management. It is now a pollen-producing male flower.

Male or Female?

I once had an avocado tree in my office. Its name was Chauncey. Chauncey was started from a seed after the fruit had provided some great guacamole. That was in Nebraska in 1976. When we moved to Virginia, Chauncey came along as a leggy one-year-old. He – actually, he-she – was my office mate for 22 years. Chauncey never got big. At 23 years old, an avocado tree can be 30 to 50 feet tall. My office was not that spacious. When I finally put Chauncey out to pasture, he was gnarly and only 5 feet tall. "Out to pasture" in this case meant withdrawing life support: water, fertilizer, and a place at the window. My office suddenly was much brighter, but I missed Chauncey. Chauncey surprised me a couple of times by producing flowers. I was not observant enough to see if those flowers did the normal avocado thing of being first female and then male. It wouldn't have mattered. There probably wasn't another avocado tree blooming within 600 miles – either to receive or provide pollen.

While some plants go to considerable lengths to avoid self-pollination and self-fertilization, others have gone entirely in the opposite direction. (Another lecture from Genetics 101 will be skipped.) The flowers of some species are designed to ensure self-pollination. Several members of the bean and grass families and wild violets and pansies have such flowers. Soybean, which is one of the world's more important crops, has quite inconspicuous flowers. One almost has to know what to look for to find them. They are small and self-contained. Their petals fully enclose the reproductive parts – both female and male. It's a shotgun wedding and a tightly closed marriage. Soybean flowers are so tightly folded up that a soybean breeder has to perform surgery before cross-pollinating two promising lines. In the emasculation surgery, which seems a little close to home right now,

The Gyroscope of Life

the pollen-producing anthers are removed before they make pollen. Then, pollen from another source is used to fertilize the now-exposed pollen-receptive surface. The offspring (seeds) from that arranged marriage will have a new mix of genes. The breeder will typically plant the offspring from the cross-pollinated soybean flowers and allow them to get back into their closed-flower, closed-marriage mating pattern. The breeder will select plants from that first generation with traits that look promising, and the seeds from those plants will be planted and the process repeated. After several generations of selecting for valuable traits, a new soybean variety may be the product.

Foregoing sex in flowering plants: Asexual reproduction

Some flowering plants have a cavalier attitude about sex. In fact, some species seem to show a preference for asexual reproduction. Willows, aspens, sumacs, and mangroves are woody species that prosper largely by making new trees asexually or clonally. In a recently discovered case, a well-known, non-woody plant has been shown to avoid sexual reproduction semi-permanently in favor of asexual reproduction. We'll get to that one after considering what may be the largest single organism in the world.

Quaking aspens are iconic trees of the western US with grayish-white bark and autumnal displays that can turn an entire mountainside golden and red. In some cases, when that mountainside turns golden, it is a single tree that is doing it. It can look like thousands of trees, because there may be that many stems or trunks involved. But a closer look sometimes reveals that those thousands of stems are all con-

Male or Female?

nected to a common root mass. They are clones from a single original tree whose roots spread and produced additional stems – sometimes called suckers. And the root lives on and on – spreading and producing more and more suckers. This way of spreading is sometimes called vegetative growth or vegetative reproduction, since sexual reproduction is not involved. The results can be mind bending.

The largest known clone of quaking aspen lives in Utah. It has about 47,000 stems or trunks, but it is just one male quaking aspen. Genetic evidence – DNA fingerprinting – shows every aspen in the clone is a perfect match for every other sample. The clone has been named Pando, from the Latin for "I spread". Pando covers over 100 acres, and he weighs in at perhaps over 6,500 tons. Pando is thought to be at least 80,000 years old. The average age of Pando's individual trunks is about 130 years. (Aspen clones can be successful in areas subject to wildfires, because, although all of the aboveground stems may be killed in an intense fire, the underground root system is not, and it can quickly produce new stems.)

While we are considering big clones and vegetative reproduction, we should mention a phenomenal fungus. (Fungi are in their own kingdom. But this fungal achievement deserves mention.) A clone of honey fungus growing in the Blue Mountains of eastern Oregon covers over three square miles (over 1,900 acres) and is estimated to be 2,400 years old and to weigh as much as 7,500 tons. The colony has been dubbed the "Humongous Fungus" (as have some other nearly-as-large fungal clones in other states). Humongous Fungus is not likely to become a tourist attraction. The honey fungus spends most of its life underground. It feeds on roots of live trees, and it spreads its black, shoe-string-like

The Gyroscope of Life

filaments through the forest floor. Occasionally, it produces aboveground, spore-bearing mushrooms. Genetic fingerprinting has clearly established that the mushrooms and the fungal filaments found throughout the colony, or clone, are from a single individual.

Dandelion is a successful species. It can be found just about everywhere in temperate regions. It is native to Europe, but it has spread from there to roadsides, gardens, and lawns all over the world. It has at least two attributes that make it so successful. One is its "parachuted" seeds that can deliver a new generation many miles downwind. A second is dandelion's ability to grow in disturbed areas like gardens, roadsides, or burned-over areas. In fact, it thrives in such conditions. It is sometimes described as a pioneer species, because it is an early settler in areas that have been disturbed.

One thing that dandelions would seem not to have going for them is sex. The majority – probably the vast majority – of dandelions that Americans see in their lawns, flower beds, and disturbed areas are reproducing asexually. Instead of relying on the good old pollination, fertilization, and seed formation sequence, many dandelions practice a sort of virgin birth – as do some animals discussed a few pages ago. The eggs in dandelion flower heads don't wait around for pollen to show up. Instead, they take the bull by the horns and produce seeds using only their own chromosomes and genes. There is some risk to this way of doing business. All of the dandelions produced are going to be just like Mom. There is no genetic diversity and environmental flexibility within the population. But clearly asexuality is working for dandelions. I can see it working every spring in my lawn and garden. It seems there can be survival value in giving up sex, as long as you can still reproduce and skirt the dictum of

Male or Female?

HR#3. It's the no-frills thing. If it's not needed, jettison it. For some virgin-birth dandelions, male sterility comes along with giving up sex. Pollen can be energetically expensive to produce. If a plant can save that energy cost, it might be able to make more or better seeds. No pollen. No frills.

Yeah, make flowers; but when?

For most flowering plants, there comes a time to show your stuff and make flowers. Especially for those plants that rely heavily on cross-pollination, there is some urgency in timing flower production, so that lots of potential sexual partners are flowering at the same time. And, especially in temperate areas, where there are distinct growing and non-growing seasons, it's important for most species to get the flowering process rolling in time to make good seeds before winter arrives. In some areas, the growing season is determined by rainfall patterns. Timing of flowering is important there also, of course.

Maybe a good strategy for a plant would be to start making flowers as soon as it can? That strategy fails, because making flowers is an expensive proposition for a plant. A young plant has business more urgent than sex on its agenda. It needs to establish roots to anchor itself and to obtain water and nutrients. It needs to make stems and leaves to harvest the Sun's energy. Only after it has attained sufficient size to support flower and seed production, should it turn its attention to sex. Essentially all plants have a genetically enforced juvenile period when they cannot flower. For perennials, especially trees, the juvenility period may be several years.

So, how does a mature plant decide when to flower in

The Gyroscope of Life

order to make good seeds before times get hard again? For some plants, especially those growing in areas subject to wet and dry seasons, the timing of flowering is determined by when it rains and then how long it takes to complete the juvenile period. Desert areas can transform from brown and gray to green and ablaze within just a few weeks after a long-delayed monsoon. So-called desert superblooms are spectacular. Seeds of numerous species that have lain dormant for years spring to life by the hundreds of thousands and complete their life cycle in the short span of time that water remains available. These seeds are not readily fooled by a shower. They have mechanisms that can, in essence, gauge how much rain has fallen, and they will not germinate until there has been sufficient rainfall.

For perennials and annuals in temperate, moderate rainfall areas, the timing of flowering needs to be tied to the calendar. Within a geographical area, each species tends to bloom at about the same time each year. There is a progression of different species blooming. In my wife's garden, hellebore plants, a.k.a., Lenten roses, flower in midwinter. Then comes snowbell, then crocus, then daffodil, then tulip, and so forth. And, in the woods, one sees a similar progression: skunk cabbage, then spring beauty, then bloodroot, then serviceberry, then redbud, then dogwood, and so forth. In the hills of Appalachia, a cold spell – often with accompanying snow – that comes when the redbuds are blooming is called a redbud winter. It can be followed by a dogwood, locust, or blackberry winter.

How does each species determine when it is time for its members to be blooming? Many plants are able to monitor their environment and use cues or clues that reveal what time of year it is. Many spring-blooming plants – crocus,

Male or Female?

daffodil, apple, peach, etc. – have been keeping track of how many cold days there have been. When they have experienced enough cold days, they will initiate flowering as soon as it finally warms up. That can lead to issues for apple and peach growers, but it generally works well for wild flowers. Flowering after exposure to a sufficient number of cold days leads to spring bloomers whose fancy lightly turns to thoughts of love after a cold winter.

The other major cue for the timing of flowering is day length, or photoperiod. For any latitude more than a few degrees north or south of the equator, day length runs through a very noticeable annual cycle. Day length is shortest at the winter solstice and longest at the summer solstice. Day lengths in between those two points follow a sine curve. This annual pattern provides a precise way of determining when to flower if a plant has some way of measuring photoperiod. Many plants have an onboard timer, a stopwatch of sorts, that monitors night length. This timing mechanism works amazingly well. Some plants that live on the equator, where annual day length varies by only a few minutes, can still rely on photoperiod to time their flowering.

In praise of flowers

Finding and photographing wild flowers became an avid avocation for me in my college years. I don't spend nearly as much time in the woods now, and a camera seldom comes along when I am out, but finding something new, or even seeing an old flower friend, still brings a rush. One of the reasons I go to the woods less often now than I did as a randy college kid is that it was often possible to talk a pretty young woman into going botanizing back in those days. My

The Gyroscope of Life

wife is still quite happy to go to the woods, but the satellite activities have changed.

Over the years, my favorite camera subjects, by far, have been the orchids. They are some of Mona's showiest flowers. One of the things that makes them so attractive – and sometimes so weird – is that they are bilaterally symmetrical. Most flowers are radially symmetrical; any line drawn through the center reveals a mirror image on either side. However, most orchids, many members of the legume family, wild violets, snapdragons, larkspurs, foxgloves, and other plants have flowers that produce a mirror image in only a single plane, i.e., they are bilaterally symmetrical. Something is inherently more attractive, more adventurous in being bilaterally symmetrical. Maybe it's because we humans are bilaterally symmetrical. One member of the legume family was scandalously, in the eyes of many naturalists, named *Clitoria* for its similarity to human anatomy. Early botanists seemed to have fun finding botanical and human anatomical similarities. The orchid family is named after the Eurasian genus *Orchis*. That name comes from the shape of the plants' paired underground tubers – "orchis" being Greek for testicle. Orchiectomy (see above) comes from the same root word.

(Here is a somewhat prurient insider joke based on Latin binomials: In both classical and ecclesiastical pronunciations of Latin, a long 'i' is pronounced like the 'i' in machine. The location of the 'i' in *Pinus*, which is the genus name for the pines, calls for it to have the long sound. Therefore, if pronounced as it would have been by Romans or Roman Catholics, *Pinus* comes off sounding like something else. The convention, however, is to give Latin binomials anglicized pronunciations. So *Pinus* is pronounce with an 'i' that

Male or Female?

sounds just like the 'i' in pine. When one looks at some of the species name for the pine genus, that makes particularly good sense. Such species include *Pinus flexilis* (limber pine), *Pinus rigida* (pitch pine), *Pinus contorta* (lodgepole pine), and *Pinus ponderosa* (ponderosa pine). But I digress (and perhaps transgress)...)

A number of orchids have used their bilateral symmetry to mimic insect pollinators. Linnaeus didn't understand the significance of such mimicry, but he commented on an orchid that produced flowers so fly-like that they were essentially indistinguishable from an actual fly. Charles Darwin understood what was going on and was fascinated by the ability of orchids to attract pollinators. The book he published right after his seminal *On the Origin of Species* was about the "various contrivances" orchids use to attract insect pollinators. What he understood clearly was that all of those frilly contrivances were not frills at all. They had a practical purpose – to get pollinators close enough to bring and take away pollen.

Studies have shown that various orchid species produce flowers that look like the female of a specific insect – a wasp, a bee, a fly, a gnat, etc. They are such good imitations that horny males are lured in for pseudocopulation (that's a technical term), during which they inadvertently pick up pollen to be carried to another pseudo-female. In most flowers, pollinators come to the flower to get pollen or nectar, but these pollinators have only sex on their soon-to-be-thwarted agenda.

In at least one case, the orchid also produces a chemical that mimics the scent of the female pheromone of the pollinator species. This, of course, further ratchets up the enticement for visiting males. The experience can be so intense

The Gyroscope of Life

(and frustrating) for the male that it ejaculates into the flower. That's clearly a waste of the insect's sperm and time, but the extra time and intimacy increase the orchid's odds for a successful pollen exchange. In another recent finding, some orchids have been shown to produce a chemical that is similar to but not identical to the female insect's pheromone. Apparently it's not that the plant is a bad chemist. Rather the males of the mimicked insect appear to be more highly attracted to the slightly strange odor. Readers are welcome to provide their own punchline. For each of these examples, the orchids, by mimicking a specific pollinator, are putting all of their eggs (or at least all of their pollen) into one insect's basket. For each different orchid species, the mimicking ruses are aimed at a particular species of insect.

Why don't orchids just make nectar the way most decent and generous flowers do? For many angiosperms, the enticements for a pollinator to visit are petal color and sweets (nectar). The colors that insects perceive, by the way, are not necessarily what we see. Insect eyes are more attuned to wavelengths that the human eye cannot see, such as ultraviolet. The UV-reflecting pigments in flowers probably look to insects like neon signs that say "land here" and "good food". Orchids may have the "land here" and "good food" markers, but many don't bother to set out candy. They rely more on the horniness of the male pollinators of a particular species. Candy is dandy, but pheromones are quicker? Nope, it just doesn't have that Ogden Nash ring.

Autobiologically, it would seem there is fitness and survival value in this frilly-flowered, one-pollinator approach. What would cause orchids to go down the path of sexual deception of a single pollinator species rather than the path of a free nectar lunch for multiple pollinators? Perhaps because

Male or Female?

it is cheaper to make weird flowers scented with pseudo-pheromones than it is to provide a banquet of sugary water. Or, perhaps, developing a relationship with a single species of pollinator ensures that pollinator will not go messing around with other flowers. After it picks up pollen from one plant (that its small brain knows was not a very responsive female wasp, bee, fly, or gnat), it will seek out another wasp, bee, fly, or gnat of its own kind. The orchid hopes it will be fooled again. Whatever the evolutionary logic, it works.

The sometimes outrageous orchids that one can see at orchid shows are not so finely tuned sex machines. That is because orchids are promiscuous. Orchid breeders can readily cross different species of orchids and get some fantastic hybrids. There is no natural selection going on for how well the flowers of these hybrids imitate pollinators. Breeders are not looking for pheromone knockoffs. The emphasis is entirely on aesthetics – or gaudiness, depending on one's perspective. In any event, many of the orchids available commercially have become biological monsters as far as Mona is concerned – not suitable for life in the wild at all. They have become domesticated (a topic for the next chapter). It's the wild ones – both literally and figuratively – that I like most.

9

Species, Evolution, and Domestication

No one definition has satisfied all naturalists; yet every naturalist knows vaguely what he means when he speaks of a species. Generally, the term includes the unknown element of a distinct act of creation.

Charles Darwin (1809 - 1882)

My evolution regarding evolution

I'm not sure when I first thought the theory of evolution was anything other than a work of the Devil. I was raised to believe all living things were God's handiwork and Adam and Eve were real. I believed Noah could sire three sons when he was 500 years old, build a boat about half again as large as a Staten Island ferry, and then herd pairs of land creatures onto that boat for a 5-month cruise. It was all in the Bible. Not to be gainsaid. I remember enjoying the snide remarks about fools who believed we were descended from monkeys or who thought that the universe could be anything other than a divine assemblage.

But perhaps because I was a pretty good student of the Bible, I began to struggle with the creation accounts in Genesis. What was the nature of the light that began to shine on the 1st day, since the Sun wasn't formed until the 4th day? Was creation a literal 6-day event where human beings were the crowning achievement? Or was creation the progression described in Genesis 2, where Adam was apparently formed before any other animals, and where Eve was the last being

Species, Evolution, and Domestication

to be created? And who were the people that Cain feared when he was expelled from God's presence for murdering his brother? There should have been only three people on Earth at that point. Who was out there for him to marry? Who was out there to inhabit the city that he built? In Genesis 6, who were the antediluvian "daughters of men" who were apparently distinct from Adam's line and who were doomed to die in the Flood?

At some age, likely in high school, I formulated a hypothesis that might resolve these seeming conflicts. They must be resolvable, because I knew the Bible was inerrant and totally self-consistent. My theory was that God created Adam from the dust of the Earth in Genesis 2, but there were already human beings on Earth. Those pre-Adamic people and many other living things would have dated from the Genesis 1 creation. Those first human beings might even have evolved into God's image under His oversight. Maybe the Genesis 1 account was the more symbolic, such that the 6 days of creation were not literal 24-hour days. With those bits of conflict-resolving logic and speculation – born out of faith in the inerrancy of scripture – came the freedom to look at evolution as a possible explanation for much of the variety of life. Perhaps both evolved life forms and created life forms exist. I might not know which are which, but it would be okay to look for patterns and clues that might support my theory. By the time I got into college biology classes, my thinking had begun to gravitate toward the notion that a lot, if not all, of life's diversity is a product of eons of natural selection – evolution.

For my baccalaureate graduation in 1967, my parents gave me two gifts. I still treasure them. One is the three-volume illustrated set of Britten and Brown, once considered the premier authority for identification of plants of the northeastern US. The other is an interesting ceramic statue about 12 inches tall. It went to graduate school with me, and it sat in my office at Virginia Tech for 34 years. It depicts a chimpanzee sitting on a stack of books – one with "Darwin" emblazoned on it. The chimpanzee is striking a Thinker-like pose, resting its chin on its left hand and gazing at a human skull in its right hand. It is a Rorschach Test of sorts. I'm not sure my parents and I saw the same thing in it, and we never talked about it. But, I can't help but feel they were telling me, "It's alright. It's okay to think for yourself." I never shared with them my evolving notions about evolution, and they never raised the topic either.

Species, taxonomy, and evolution

Aristotle, the great philosopher and naturalist, introduced the concept of species. He considered species to be distinctive, unchanging groups of organisms – unchanged since their creation. His view of species being fixed and stable entities was stamped on western naturalist thought for two millennia. When Linnaeus, the father of taxonomy, began his task of naming and categorizing living things in the 18th century, he chose species as his basic unit of classification. This made great sense, because he accepted Aristotle's notion that animal and plant species are divinely fixed – the same yesterday, today, and tomorrow. One could build a naming system solidly on such a foundation.

Besides naming species, Linnaeus formed other taxonomic categories for plants and animals. His groupings graduated from species, to genus, to order, to class, and then to kingdom. That system was expanded by later taxonomists into

Species, Evolution, and Domestication

a seven-layer hierarchy, and, in the 21st century, domains were added atop kingdom for an even more inclusive grouping. For Linnaeus, the categories above the species level did not imply any kinship. They were merely pigeon holes into which he could place species with similar features – learning aids for appreciating the multitude of God's species. In his mind (and the minds of all naturalists in that time), the members of a species were kin only to one another – not to any other species. Every species was an immutable, stable-forever life form, and each species still looked just like its originally created ancestor. But that long-held notion did not stand up to critical examination by some famous 19th century thinkers.

In his later studies, Linnaeus discovered hybridization between plant species. This was quite a revelation for him. He said in 1760, "It is impossible to doubt that there are new species produced by hybrid generation." Coming from no less a person than Linnaeus, this was a significant blow to the 2,000-year-old view that species were divine and immutable creations. It opened the door so that just a glint showed through.

And then 99 years later, along came Charles Darwin. *On the Origin of Species* (1859) blew the doors off and introduced an entirely new take on species. They were no longer seen as the immutable products of creation. Rather, they came to be viewed as the ever-changing products of eons of evolution. Perhaps not coincidentally, in Darwin's time, naturalists were evolving from being natural historians into becoming biologists and scientists. The move from seeing living things as divine creations to seeing them as biological products meant species were no longer just to be catalogued and named. They were now legitimate subjects for investiga-

The Gyroscope of Life

tions into kinships and causality. But now biologists needed a new definition for species.

A standard modern definition for species is this: groups of organisms with shared traits and so genetically similar that they can produce fertile offspring. But, for many biologists and taxonomists, that definition fails to satisfy. It implicitly supposes that a unique set of species-identifying traits can be found and agreed upon. Biologists have a long history of disagreeing on lots of things, to include what constitutes any given species. The "species problem" is a recognized area of philosophical and technical debate among biologists. Some odd biological phenomena fuel that debate.

Here's a thought experiment that illustrates part of the species problem. Imagine three groups of crayfish, a.k.a., crawdads, living in streams in three adjacent Appalachian hollows, or hollers as they are known back in the hollers. The crawdads from holler A occasionally wander downstream to the confluence of the stream from holler B, wander up into holler B, and mate with crawdads there, and vice versa. The same kind of hanky-panky goes on between hollers B and C. When their crawdads get together, baby crawdads happen. But, when crawdads from holler A and C get together, there is no chemistry between them. They are not compatible sexually, although they look alike and although they both can successfully mate with crawdads from holler B. So, by the standard definition, the crawdads in holler A are a different species than those in holler C, because the two cannot mate and produce fertile offspring. That would be okay, but it leaves the crawdads in holler B in a puzzling situation. To what species do they belong? This turns out to be more than a thought experiment. A few documented cases in the animal kingdom and at least one in the plant kingdom sound

Species, Evolution, and Domestication

much like the crawdad hollers story – except with more hollers or populations involved. The resulting conundrum is called a ring species, just one of the problems in defining species.

And there is the semi-embarrassing case of the fungi imperfecti, or imperfect fungi. They include fungi that cause athlete's foot and others that give Roquefort and Camembert cheeses their distinctive aromas. (Maybe that connection explains the smelly foot odor some associate with those cheeses?) Imperfect fungi do not reproduce sexually. Sexual reproduction in perfect fungi, which include the majority of fungi, involves making mushrooms or other spore-bearing structures. The imperfect fungi produce spores only by asexual means and without showy spore-making structures. So these atypical fungi were described as solely asexual species – even though a species is partially defined by its ability to reproduce sexually – and named with a proper Latin binomial. Only after many years did mycologists discover that some of the imperfect fungi are long-lived asexual stages of already-known fungi, and, under the right conditions, they will also reproduce sexually. With the advent of DNA sequencing, the sexual forms of many more imperfect fungi were discovered. These were all cases where the same species had been given two different names.

A somewhat cynical amphibian taxonomist once told me, "A species is what you can get your colleagues to accept". There is more than an element of truth to that. There is a good bit of subjectivity (and artistry) in taxonomy. According to the rules that Linnaeus set up, if a taxonomist finds an organism that has not been previously described and she or he develops a good description of it, she or he gets to put a name on it and publish the finding. But taxonomists differ

The Gyroscope of Life

in their approach to delineating species. They often speak of one another – and not always in a kind way – as lumpers or splitters. Darwin may have coined the terms. In a letter about defining species boundaries, he wrote to a colleague that "It is good to have hair-splitters and lumpers." Lumpers tend to use a smaller or more general set of traits to identify a species. They can therefore find larger, more diverse collections of organisms to lump into a species as they define it. Splitters, on the other hand, develop a longer set of identifying traits, one that can exclude some or many of the organisms fitting into a lumper's definition. Some taxonomists take a middle ground, naming a different form not a new species but a subspecies, variety, landrace, or ecotype.

When we step back and think about what leads up to the naming of a new species, it is probably inevitable that species are hard to nail down descriptively and taxonomically. A species should be difficult to circumscribe. Evolution is an ongoing process. All species alive today have arrived at whatever point they are on their evolutionary journeys at the same time that *Homo sapiens* has arrived at a stage where we have a penchant for naming things. But every species is in evolutionary flux. Taxonomists are, in essence, shooting at a moving target, and, rather than presenting a target that offers a distinct center of mass, the target is more often amoeba-like with projections sticking out here and there.

You probably know the classical story from India about six blind men who come upon an elephant. One touches and explores the trunk. Another does the same with a leg. Another with an ear. Another with a tusk. Another touches the elephant's flank. And the sixth grasps the tail. When they compare their experiences, they conclude they were dealing with six different animals. Most taxonomists are not blind,

Species, Evolution, and Domestication

but they may sometimes overemphasize the uniqueness of a population found at one locale. Accordingly, they might become splitters.

Until recently, decisions on what traits were used to group organisms into a species were often made after collecting and studying many individuals. Taxonomists would study the details of their organisms' shape and makeup. They would begin to see patterns – recurring traits – unique to their collection. Those became the basis for identifying their collection as belonging to a new species. It makes great sense, but it is subject to the elephant effect. Problems can arise from odd-looking subsets of a species that technically remain within the gene pool because they can breed with the mainline members of the species. Within the last few years, though, taxonomists and systematists have begun using DNA sequencing to identify closely related individuals – so closely related that they must be in the same gene pool and therefore in the same species. The new approach to finding genetic connections is moving taxonomy and the definition of species into new territory.

A taxonomic case history, and taxonomists' new tool

I worked for many years on the culture and uses of switchgrass. The genus to which switchgrass belongs (*Panicum*) is quite large, with almost 500 species worldwide. All 500 *Panicum* species were placed in that genus because they share a distinctive slender, multi-branched seed head. The great similarity of seed heads across all 500 species caused taxonomists to suppose all were fairly closely related – not just in body type, but genetically and evolutionarily. If the

The Gyroscope of Life

taxonomists were right, all 500 could trace their evolutionary ancestry back to a single original *Panicum* – the Adam and Eve *Panicum*. All present-day *Panicum*s would be the umpteenth-great-grandchildren of that first *Panicum*, and all would be cousins, albeit with some being many, many times removed.

But evidence from DNA sequencing studies now shows that some *Panicum* species are not closely related. In fact, some are more closely related to other genera of grasses than they are to other *Panicum*s. The new evidence shows that slender, multi-branched seed heads – the trait that made all of these grasses *Panicum*s – does not provide clear evidence of kinship. Many of those 500 *Panicum*s are in taxonomic transition now – eventually headed off into other genera.

The DNA-sequencing technology that proved the *Panicum* genus is a mishmash of not-so-closely-related grasses is being used to provide solid evidence about evolutionary kinships among all living things. Instead of making comparisons at a visible or physical level, for example, delicate, multi-branched seed heads, the new breed of taxonomists looks at the DNA of specific genes or entire gene sets. For these studies, DNA is isolated from organisms, and the degree of similarity between their DNA, or genes, is determined. A perfect match would support the likelihood that two organisms are the same individual, as in the case of our 100-acre aspen clone called Pando. A 99.9% match could mean we are looking at another organism from the same species. A 99% match would suggest the two organisms may be closely enough related to be in the same genus or family. (Human and chimpanzee genes are about 99% matches. We are classified in the same family: Hominidae). And so forth. The logic is quite straightforward: the more alike the genes

Species, Evolution, and Domestication

of two organisms are, the more closely related the two are evolutionarily. Related means the same thing for plants and animals that it does for us. Relatives share an ancestor. The degree of genetic similarity is a good indicator of how far back into evolutionary time – how far down the branches of the Tree of Life – we might need to go to find that shared ancestor.

DNA sequencing has rapidly become the method of choice for making taxonomic decisions and refining the shape of the Tree of Life. Such data are helping to bring order to taxonomy and to create a true or natural picture of the relationships among living things. But many traditional taxonomic apple carts are being upset in the process. Some of those were discussed earlier in consideration of how many kingdoms and domains there might be in the biological world. DNA sequencing showed, for example, that fungi are more closely related to animals than plants. So they ended up in a totally new apple cart – their own fungal kingdom. Their branch on the Tree of Life is now a shoot off of the eukaryotic branch also occupied by animals (and plants constitute a third shoot off that same branch).

Evolution: An inevitable outcome of changeable species

Naturalists such as Charles Darwin, who traveled widely to observe and collect specimens saw things that stay-at-home naturalists would not. The observations that Darwin made in his early travels perhaps flitted around in his mind, looking for a perch. But as he observed more and more living things, forms with similarities would perhaps have begun to settle into different mental pigeon holes. I can imagine his

The Gyroscope of Life

mind eventually being able to produce parades of organisms that lined up by type and then marched by him in an order that suggested connections and even kinships. His brilliant mind perhaps didn't have to make too great an autobiological leap, then, to envision a process whereby such a menagerie could have been formed. That process, of course, was evolution, which explains the origin of species and much more.

Natural selection and evolution of species happen – naturally enough – at the species level, but, over evolutionary time, natural selection results in differences recognized at higher taxonomic levels. For example, once closely-related land plants went their separate ways evolutionarily hundreds of millions of years ago and trudged down different evolutionary paths to become today's organ pipe cacti and redwood trees: huge changes that all happened just one step and one species at a time. The fork in the path leading to redwoods or to cacti came pretty early – perhaps shortly after seeds first evolved in fern-like plants about 400 mya. Cacti are flowering plants, and redwoods are conifers, but both rely on seeds during sexual reproduction. We can suppose, then, that their lineages parted ways after seeds evolved. Some of those neophyte seed-bearing plants went on to become conifers. Others diverged and went on to become angiosperms. Over time, some early conifers separated evolutionarily from the pack and began to acquire characteristics we associate with redwoods. Similarly, some flower-bearing plants in drier environments began evolving toward looking more and more like barrels with spines. The nitty-gritty of all those changes was done as natural selection gradually remodeled existing species – remodeling them until they no longer looked like the organisms they once were. Cacti and

Species, Evolution, and Domestication

redwoods are now so distantly related that they often appear together only at a phylum level. Yet, we know their ancestors were once – before evolution did its thing – much more closely related. In fact, they have a common ancestor way back there when seeds were becoming the way to go on land.

Natural selection and evolution are not just options or possibilities; they are inevitabilities when species are mutable. Species are not the unchanging, fixed beings they were once thought to be. They have plasticity, shifting in shape and other characteristics over time and space. We now know (but Darwin didn't) that species' plasticity arises from mutation of genes and from mixing of genes during sexual reproduction. Several mechanisms for modifying or mutating DNA and for creating new combinations of genes in succeeding generations are now well-understood. The wizard behind the curtain has been revealed. Biologists understand how and why organisms differ genetically from their parents, but we will not develop that theme here. Suffice it to say that genetic change is natural and inevitable, and that means species will – even must – evolve.

Evolutionary time scales

How long does it take to see the results of evolution? In a petri dish, evolution can occur in time frames measured in days. In such a demonstration, a species or strain of bacteria is allowed to grow until it covers the surface of the dish. Then a poison, such as an antibiotic, is applied. Typically, the bacteria die off rapidly. But sometimes a little spot of life develops in what is otherwise a field of bacterial corpses, and, with time, that spot grows and might again cover the whole dish. Adding more poison has no effect. Genetic fingerprint-

The Gyroscope of Life

ing of the original strain and the surviving strain show that they are identical except for one gene. The logical explanation is that the bacteria growing in our poisoned microcosm have developed resistance to the toxin. They have evolved.

That same sequence has been observed in the macrocosm of real life and in time frames measured in years. Since the 1940s, medical science has been using antibiotics to fight infections and diseases. Antibiotics have been a boon, because they kill bacteria that cause serious or fatal illnesses, and they stop life-threatening bacterial infections. But some infection- and disease-causing bacteria have evolved in less than a human lifetime into forms that are now resistant to standard antibiotics. This story is well documented. Many factors can be at play, but the ultimate cause (culprit, if you must) is mutation and natural selection – evolution. The bad-acting bacteria have evolved to fit an antibiotic-laced environment. They're just obeying House Rule #1, evolving to fit their altered environment. Something similar is happening in the weed world, with several noxious weeds developing resistance to herbicides.

Table 9.1, which is reworked from Table 4.4, keys on some evolutionary milestones. Studying it gives us some notion of the pace at which evolution has unfolded. At the domain or kingdom level, evolution is glacially slow. The putative earliest life forms eventually gave rise to bacteria and archaea, and, for close to 2 billion years, natural selection worked exclusively within those two domains. Photosynthesis evolved within the bacteria, and by a billion years later, the oxygen being generated in photosynthesis built up to the point that it became toxic. That had disastrous consequences for many, maybe most of the organisms then alive. But Mona and natural selection developed new species that could

Species, Evolution, and Domestication

live in environments laced with oxygen. New oxygen-tolerating bacteria and archaea became well-established across a diverse range of environments. One supposes that, along the way, lots of adaptive and protective mechanisms had to be developed by natural selection's trial and error method. But after 4 billion years or so, today's bacteria and archaea clearly have quite a repertoire of adaptive and protective mechanisms. Unfortunately for us, some of those mechanisms can apparently be deployed – maybe just with minor modification – to defeat antibiotics.

Table 9.1. Key events in biology from life's first appearance to modern times. The question marks denote times that are still open to considerable debate.

Billions of Years Ago	Event or Stage
4.3?-3.8	Life appears; Bacteria and Archaea evolve
3.5	Photosynthetic organisms appear
2.4	The Oxygen Crisis, massive extinctions
2.0?	Single-celled eukaryotic/higher life forms appear
1.2	Sexual processes evolve (for single cells)
1.0?	Fungi branch off of early animal forms
1.0?-0.8	Multicellular eukaryotic/higher life forms appear
0.56	Cambrian Explosion (of aquatic animals)
0.4	Life on land develops and evolves
0.13	First flowering plants appear
0.00025	*Homo sapiens* appears
0.0000023	Aristotle: species are unchanging creations
0.00000015	Charles Darwin: species evolve

The Gyroscope of Life

What about the evolutionary journey of higher life forms: plants, animals, and fungi – the eukaryotes? The first eukaryotes appeared at least 2 bya, and some evidence suggests they may have arisen a half billion years earlier. But Life 2.0 did not race down the evolutionary yellow brick road. In the most generous analysis, it took at least 800 million years for eukaryotes to come up with sex. It may have taken another 200 to 400 million years for them to figure out how to play together as more than single cells. Multicellularity was a huge step to take evolutionarily. It required more than just developing cellular glue to hold cells together. If cells are going to congregate and act as a single organism, they must be able to: 1) coordinate activities, 2) develop cellular specializations to carry out key functions (like keeping oneself internally wet on dry land), 3) have every cell carry (in its genes) a complete set of instructions for how to assemble the entire cellular congregation, and 4) have each cell utilize only the portion of the instructions pertaining to its particular function. Given the logistics and logic to such a huge information technology undertaking, it is perhaps not surprising that it took eukaryotes 1 to 1.2 billion years to come up with all the requisite particulars.

The systematists who study the evolutionary development and diversification of organisms using DNA evidence, suspect plants, animals, and fungi solved the IT problems of multicellularity separately. Single-celled plants and animals evolved from lineages of primitive single-celled eukaryotes, and single-celled fungi then branched off from animals while all three kingdoms were still single-celled. Then each kingdom set about figuring out how to be multicellular, with the fungi perhaps being first. But, once the IT problems were solved, evolution seemed to pick up the pace. The Cambrian

Species, Evolution, and Domestication

Explosion of marine animals came just 240 million years after the first multicellular forms evolved. Some of those animals and many plants learned how to move out onto the land in another 160 million years. Or maybe less. Numbers in such cases are merely a reflection of how old the fossils are that exhibit a new characteristic or behavior. We can't really know that was the first ancient form to achieve that milepost. It's just the oldest that we know of. You can expect many of the numbers you've just read to be outdated soon, maybe even before you've read this.

Regardless of the pace and timing, evolution and natural selection worked on species individually to produce the collective that is life. Living things have established footholds and now thrive in environments as diverse as rainforests and deserts, equatorial savannahs and polar tundra, cool tidal pools and boiling hot springs, Mediterranean hillsides and Alpine peaks, mountain freshets and super salty seas. Life can be found in all of these and many more environments, ranging from those that would seem life-friendly to those that are definitely life-challenging. That has happened because natural selection, when layered on top of mutable genes, has produced species that can meet the challenges and enjoy the advantages each environment presents.

Maybe it will help to say it this way: Natural selection is a process. Evolution is the outcome. Natural selection is the driving force that approves or rejects genetic variations that inevitably arise in a species. If a different genetic version helps the species better adhere to Mona's House Rules, it will become the new normal for the species. The natural selection of such genetic changes over time results in a species' evolution and, over even greater periods of time, in a Tree of Life.

The Gyroscope of Life

Artificial selection leads to domestication

Natural selection is a no-nonsense, make-or-break system for evolutionarily honing the fitness of a species. It is "natural" selection because nature, or the natural environment, is acting as judge, jury, and executioner in this life-or-death process. By contrast, unnatural, or artificial, selection processes have been in play since sometime after *Homo sapiens* came on the scene. Human beings consciously or unconsciously favor the expression and propagation of certain traits in some species. Over time, this artificial selection has resulted in domestication of both animals and plants. These species are considered domesticated because they are typically dependent on humans for their continued survival. From Mona's standpoint, they would be considered artifacts, even biological monsters – unsuitable forms that natural selection would never have produced. Some examples may help to make the point.

Man's best friend is descended from wolves. DNA sequencing shows that today's dogs separated from the gray wolf line 20,000 to 40,000 years ago. Gray wolves are finely honed predators and able to thrive in a variety of environments – clear winners in the natural selection lottery. But wolves do not make easy-going pets, even when raised from puppyhood. The same was apparently not true for the proto-dogs. While hanging around and interacting with *Homo sapiens* for 15,000 years or more, these erstwhile wolves have lost many of their wolf-like traits. Today's dogs come in a remarkable range of sizes and shapes. Some still look wolf-like. Others are bigger than wolves but have few lupine features physically or temperamentally. And then there are dog breeds where centuries of human-supervised crossings

Species, Evolution, and Domestication

have produced animals unfit for survival in any environment except the human lap. Could a pack of Chihuahuas run down and kill an elk? Absolutely not. What if a rabbit were their prey? Probably not. Dogs are domesticated animals because, over the years, humans have selected and bred for traits that made them good pets or good working dogs but left them without traits to fit wild/natural environments. See House Rule #1. Many breeds of dog simply would not survive without human intervention. So we describe them as domesticated.

Interestingly, as diverse as dogs are, from Chihuahuas to Great Pyrenees, they are all considered to be in the same species. Their antics when a female is in heat bears that out. Some very different looking and different size dogs can get together to produce a litter; they can produce fertile offspring, a chief criterion for identifying those in the same species. And their DNA shows that dogs of all breeds are not that far apart genetically. Relatively few genes have been selected for to arrive at all that variety, with more than 300 recognized breeds worldwide.

In the mid-20th century, Russian scientists began a still-running experiment on fox domestication. Foxes, as cute as they are, do not make good pets. They have not been domesticated historically. In fact, they are untamable according to most who have tried to adopt young kits. The experiment, done on a Russian fox farm, relied on (artificially) selecting and breeding "friendly" foxes – those that best tolerated human interaction. (The foxes that behaved around humans the way most foxes do – showing fear or aggressiveness – were turned into fur coats.) During 15 or 20 generations of breeding friendly foxes with one another, the offspring became more and more dog-like in their behavior. They

The Gyroscope of Life

welcomed – even begged for – human attention. Surprisingly, they also developed floppy ears, curly tails, shorter and broader skulls, shorter legs, different coat colors, etc. After 60 years of selection just for friendliness, they were much more like dogs than some dogs. Instead of being turned into garments, these tame, i.e., domesticated, foxes are now being sold as pets.

Plants can be domesticated also. In fact, the list of plants whose evolution has been co-opted by humans' artificial selection is longer than that for domesticated animals. Most of our major crops are so far removed from their wild state that they would likely disappear if humans were to disappear. It is hard to imagine corn surviving for long in Iowa if farmers were not there to clear the land, prepare the soil, and carefully plant seeds that had been safely stored over winter.

If dogs demonstrate the dramatic consequences and outlandish outcomes of artificial selection, the plant kingdom equivalent has to be cole crops. Cole crops include broccoli, Brussels sprouts, cabbage, cauliflower, collards, gai lan (a.k.a. Chinese broccoli or Chinese kale), kale, kohlrabi, and savoy. All have been selected – by humans, not Mona – for traits that make each of them desirable as food. That may not be surprising, but you may be surprised to know that all nine of the cole crops, as diverse as they are, are still considered to be the same species. In fact, DNA sequencing shows they are all in a single species of the mustard family. Tellingly, one more member of the species, wild cabbage, occurs in the wild in the Mediterranean region.

Autobiologically, here's what we might suppose happened leading to cole crops' domestication. Wild cabbage makes pretty decent greens when cooked; so it likely was identified by pre-agricultural hunter-gatherers as food. They would col-

Species, Evolution, and Domestication

lect it wherever they found it, take it home, cook it, and eat it. Eventually, though, some very smart persons (I'd bet they were from the distaff side of their communities) realized that seeds of wild cabbage could be collected and used to produce food right there at home. That is when the plant started down the road to domestication, as soon as it was brought into cultivation and no longer under the sole supervision of Mona. It was now subjected consciously or unconsciously to selection by humans, not natural selection. Over the course of many human generations, some of these cultivated plants developed peculiar vegetative or reproductive traits. Those spontaneous changes were caused by mutations of just a few genes – so few that the very distinct-looking plants have remained virtually identical under DNA sequencing. Those that became kale were selected (seeds saved and replanted) because they produced bigger stems with edible leaves throughout the growing season. Others were propagated because they began showing an ability to produce shorter stems and eventually a tightly wrapped cluster of leaves – a cabbage head. Some produced smaller heads clustered along a longer stem and became Brussel sprouts. Others evolved under human supervision into kohlrabi, as they produced – over long periods of human selection – an increasingly swollen, edible stem. Some produced clusters of edible flower buds at the top of their stems. The clusters were dense and non-green (cauliflower) or a bit more open and green (broccoli). Each form is a product of human tinkering with the rules for natural selection and survival. Each form is domesticated, no longer able to survive on its own, although wild cabbage still does.

That last sentence suggests that the cole crops struck a Faustian bargain. They are now found growing all over the

The Gyroscope of Life

world, but, wherever they occur, it is only because they are being grown (and mollycoddled) by their human masters. A similar story can be told for many other crops: wheat, oats, barley, rice, soybean, etc. Each arose from a wild species (some of which are still present in their region of origin), was placed under cultivation, and then "evolved" under human selection into the crop it is today. In the process, they became better and better food sources for us, and they became less and less able to obey HR#1 (at least as HR#1 applies to fitting a natural environment).

Corn, or maize, presents a challenge for those who study the history of plant domestication. No wild plant looks remotely like corn. No grasses (corn is a grass) produce seeds on a shuck-covered ear with closely packed rows of seeds. We know corn is from the New World. Native Americans introduced it to Europeans, along with tobacco, squash, beans, tomatoes, peppers, potatoes, tobacco, sunflower, and more. Ethnobotanists, the people who study the relationships between plants and human cultures, have determined that corn's point of origin was in Central America, likely in modern day Mexico. But no corny-looking plants occur in the Mexican flora today. One grass in their flora is teosinte. DNA-sequencing data and other lines of evidence suggest teosinte could have been a starting point for what became corn. If so, it required a major piece of re-engineering by human/artificial selection.

Early Mesoamericans presumably found mutants or hybrids of teosinte that possessed corn-like traits and then selectively propagated those in their home gardens. They then repeated the process with the new lines, saving and propagating seeds from individuals that, by mutation or hybridization, were more corn-like. Of course, our heroes

Species, Evolution, and Domestication

didn't know what corn was supposed to look like. They were merely selecting for traits that made teosinte a better source of food. Over hundreds of human generations, that expanding pool of human-selected traits gradually came to look more and more like today's corn. The result of this artificial selection was entirely arbitrary from Mona's standpoint. She, or natural selection, would never have produced such a child. It is telling that none of those intermediate forms have come down to us growing in the wild; they were already not able to follow the House Rules without human assistance. But for the Mesoamericans (and eventually for native peoples in North and South America), corn became the Corn Mother – one of the famous Three Sisters (corn, squash, and beans). We call it *Zea mays* as though it is a naturally selected and evolved species, but it and a number of other domesticated plants – and dogs – should probably have an asterisk after their names to distinguish them from Mona's naturally evolved children.

(Here is a party trick for the next time you are eating corn-on-the-cob. Bet anyone sitting near you that the ear they are about to pick up will have an even number of kernel rows. The rows are the more or less straight lines, or columns, of kernels running along the length of the ear. You'll win your bet essentially all the time, if you count the rows somewhere near the middle of the ear. The rows can get a little uneven with some fading out near either end of the ear. Your bet is pretty much a sure thing because of the way an ear of corn develops. That includes the formation of tissues running along the length of the developing ear. Those linear tissues ultimately give rise to kernels, but each line gives rise to a double row of kernels. A typical ear of sweet corn will have eight to twelve kernel-forming lines, so it will produce

The Gyroscope of Life

16 to 24 – but always an even number – rows of kernels by the time you chomp into it.)

Key crop traits selected "accidentally" by prehistoric agriculturalists

Artificial selection today is a conscious, directional process. Plant and animal breeders know what traits they want to introduce or improve: leaner meat, greater disease resistance, crisper apples, higher yields, greater heat tolerance, and so forth. The target trait is tested in each generation or line, and the better individuals will be chosen for further breeding. Over time, the trait moves in a direction determined by the breeder's selection process. But, with a bit of autobiologic, we can suppose that some crop traits essential to plant agriculture were selected for and refined long before anyone drew a paycheck as a plant breeder. The process was somewhere between natural and artificial selection in the sense that these traits were being selected for but only by accident. It was an entirely unconscious, directionless process – a natural outcome of early human interactions with plants. It dates back to the time when humans were, in essence, just part of the natural landscape. Here are several examples.

Plants make seeds and often help those seeds disperse so they can produce more of the same species. See House Rule #3. Some plants produce edible fruits that attract animals that eat the fruit and spread the seeds – often with a shot of organic fertilizer. (Speaking of shot – some plants literally shoot their seeds into the environment. The squirting cucumber (not edible) can build up enough pressure to spurt its seeds 10 feet or more. The dynamite tree, which is native to tropical portions of the New World, has explosive fruits

Species, Evolution, and Domestication

that can fling seeds 150 feet or more. It got its name from the sound of the exploding fruits.) Tumbleweeds, which include plants from several species, release their seeds after the parent plant dies, breaks off, and tumbles tens or even hundreds of miles. Other plants do not go to such lengths, but essentially all have mechanisms for shedding their seeds so that they get to the ground. The general term for seed release is shattering.

In an ideal crop, the plants will make grain and dry down without shattering. Then their seeds can be easily harvested by hand or machine. Our major grain crops don't shatter in spite of the fact that their wild progenitors did. We can probably thank the first people who adopted early crops. They unconsciously selected for non-shattering plants. As they walked through a patch of primitive wheat or rye, they would have collected the seeds that were still on plants. Unknowingly, they were selecting from plants that had a reduced tendency to shatter. They took those seeds home and ate some of them, but, more importantly, they planted some. That act of faith, taking food out of their mouths and putting it into the ground, ensured the survival of plants that didn't release their seeds so readily. The process was repeated, and some of the next generation of seeds were saved for planting. Then, there was another round of unintentional selection for less shattering. And another. With time, the umpteenth artificially selected generation held onto essentially all its seeds. The same serendipity likely happened in multiple places, at multiple times, and with multiple species destined to be our non-shattering seed and grain crops.

Dormancy of seeds is another trait that would make agriculture difficult. The seeds of many wild plants often fail to germinate after they land in a suitable soil. They are blocked

The Gyroscope of Life

from germinating by various dormancy mechanisms. Some seeds have a hard seed coat that will not allow water to penetrate, but, as they lie in the soil, fungi and other soil organisms will eventually break down the seed coat and allow germination to begin. Some seeds are unable to germinate until they are exposed to cool temperatures or until they sit around for a year or more.

Dormancy is desirable from Mona's standpoint. Seeds that begin to germinate as soon as they hit moist soil may produce seedlings that don't have time to flower and make more seeds before the growing season ends. There is survival value in producing seeds that might lie dormant for years. But none of those reasons for dormancy fit readily into farmers' plans. Farmers want seeds that will begin to grow quickly after planting. They can probably thank those first agriculturalists for making that possible. A bit of autobiologic suggests those proto-agriculturalists did not intentionally select for non-dormancy. They planted seeds and then harvested from the plants that came up, i.e., those whose seeds were not dormant. Over time, a larger percentage of the seeds planted would come up, because those first farmers were, without realizing it, selecting for non-dormancy. This unconscious, unintentional selection is the likely prehistoric backstory that explains why most of our major crops do not have seed dormancy issues.

A third feature of crops – besides non-shattering and non-dormant seeds – is essential to modern, large-scale agriculture. It probably played a significant role in early agriculture as well, and it also was likely selected for unconsciously by prehistoric agriculturalists. It has to do with when crops mature and are ready to harvest. An ideal crop in modern, large-scale agriculture will grow for as much of the growing

Species, Evolution, and Domestication

season as it can, fairly quickly wrap up its business, and then die. Crops such as field corn (but not sweet corn), wheat, and soybeans cannot be harvested when they are alive and the grain is still wet. Rather, they must dry down (without shattering) so the grain can be readily harvested and then safely stored. Mona has a mechanism to accomplish that sequence. Its technical name is monocarpic senescence. Let's just call it programmed death – PD for short.

Mona didn't come up with PD with farmers in mind. It is a strategy that works to produce more and healthier seeds for the next generation. Mona is all about future generations. The PD phenomenon occurs in many annual plants, plants that live for only one growing season and that produce seeds to carry over until the next growing season. Plants that exhibit PD shut down as they are finishing up the process of making seeds. But the shutdown is orderly. As it approaches its death, the plant mobilizes many of the constituents in its roots, stems, and leaves and sends them to its seeds. It can produce more and bigger seeds as a result. This works well from Mona's standpoint, and it's super as far as farmers are concerned.

But Mona didn't set up PD so that the plants of an annual species would all die at the same time. Some probably dropped (or shot) their seeds by mid-summer, while others stretched their growing period out till early fall. That poses no problem for a wild plant. In fact, it might even have survival value. The mid-summer seeds could be insurance. They could be the only seeds produced in a dry year or in a year when the first frost is super early. Presumably different genes or different mutants of a gene cause the PD to be smeared out over the end of the growing season.

Smeared-out maturity and harvest dates are not good

The Gyroscope of Life

for agriculturalists – especially modern, big-scale farmers. Even for prehistoric agriculturalists, it would have been more convenient to be able to harvest a small planting all at one time, rather than having to go back repeatedly to catch each plant at its individually determined maturity. One can suppose that early agriculturists may have ignored the first few plants to go through PD and then mature, if they represented a small fraction of their proto-gardens. They would have harvested heavily as the bulk of the plants matured. And then they might have left the last few seeds to the birds. This pragmatic method would have unwittingly resulted in more uniform maturities for the seeds left over to plant next year. And the same unconscious, artificial selection would take place again and again; or so autobiological thinking would suggest. With time, the population of the species they were working with would become more and more synchronized for their time of PD and harvest – and more and more domesticated. Convenient for harvesting by hand. Absolutely essential for harvesting by machine. And, again, we can probably thank the women who were the first farmers 10 to 12 millennia ago for their artificial selection leading to more uniform maturity for our crops. They likely did a lot of heavy lifting in developing crops that would be well-suited for modern agriculture.

A crop ideal that is a no-brainer – and a non-starter

Being able to make both more and bigger seeds would be ideal for a crop. Did early agriculturalists move plants in that direction by artificial selection? A bit more autobiologic suggests maybe not. Natural selection might be expected to

Species, Evolution, and Domestication

select for those traits as well. After all, making more and bigger seeds should improve a species' chances for survival. They could provide a big leg-up in complying with Mona's HR#3 about making babies and passing on DNA.

Mona does, indeed, favor the survival of some plants that make many seeds. Many wild orchids produce seeds by the hundreds of thousands. She also has plants that produce huge seeds. The biggest seeds in the world are produced by coconut trees. But Mona does not seem to have found a way for a plant to produce both many and large seeds. That would seem to create too much of a logistical strain on the parent plant. Seeds are expensive to produce from a plant's standpoint. Parent plants must provide all of the materials required to make seeds. The bigger the seeds, the more goodies the plant must generate and/or deprive itself of. Those orchid seeds produced by the hundreds of thousands are dust-size and carry nothing onboard in the way of nutrients or energy supplies. They are cheap to produce from an orchid plant's standpoint. That's why they can be produced in such numbers. Conversely, coconut trees must grow for many years until they are large enough to support the growth of those huge seeds, and it can take several years for a single coconut seed to develop to maturity. It's a no-free-lunch thing.

Modern plant breeders have also found that there is no free lunch. When they introduce traits that produce more seeds on a plant, they generally see smaller seeds produced. When they try to increase seed size, seed numbers are reduced. Corn breeders have worked hard to increase the number and size of corn kernels, and they have achieved some success. But, they have discovered there was a tradeoff. When they looked more closely at their more-and-bigger

The Gyroscope of Life

kernels, they saw a reduction in protein content. Corn kernels are mostly starch. The amount of protein is relatively small but important nutritionally. Bigger corn kernels were being made only because the plants were diverting their efforts from making protein into making more starch. Grain yields were going up, but grain quality was going down. No free lunch. That's the sort of calculus that plants must do, and it seemingly wouldn't have allowed prehistoric agriculturalists to come up with domesticated crops that could produce both more and bigger seeds.

But, again on the positive side, prehistoric artificial selection and domestication undoubtedly had major, positive (from a human perspective) impacts on the species that we have adopted and adapted as our crops. They are no longer able to survive in the wild, but they serve us well.

Vegetative propagation of crops (stopping evolution in its tracks)

We humans have taken great advantage of the mutability of species, using it to lead many of them down the garden path – both literally and figuratively – to domestication. It's a time-honored, if not Mona-honored, practice. And we continue to do so for new crops, such as switchgrass, that have just been adopted from the wild. We can suppose that switchgrass and other newly discovered proto-crops will look quite different in a century or two. The changes will come much faster than they did 12 millennia ago. We now know what traits to look for, and we now understand the genetic underpinnings of those traits. The path to domestication via human-directed evolution is now shorter and better marked.

But sometimes we want crop evolution to stop. These are

Species, Evolution, and Domestication

situations where some member of a crop species has arrived at a particularly desirable form and where a next generation would be impossible or less desirable. So, for example, we have seedless grapes, bananas, and oranges. The seedless condition can arise naturally by mutation of genes affecting seed development, and it can be a desirable trait, but how can we grow more seedless plants when there are no seeds? Instead of relying on propagation by seeds, which our otherwise desirable plants don't have, we use vegetative, or asexual, propagation. Cuttings from seedless grape vines and orange trees can be rooted and develop into more seedless grape vines and orange trees. Sprouts or offsets from the base of banana plants can be transplanted to produce a new crop. The resulting plants are clones, exact copies genetically of the vine, tree, or banana plant from which they were produced. In a sense, evolution has stopped for them.

Making seedless watermelons is more involved. Those edible wonders are the product of some genetic sleight-of-hand. Watermelon breeders have learned how to cross two different lines of watermelon such that the resultant seeds produce only seedless melon vines. The same seedless result happens every time those two varieties are crossed. So, seedless watermelons are based on sexual propagation, but one that produces a seedless evolutionary dead end. And that is the idea – to produce plants that are just like their predecessors, stopping the evolutionary wheel from turning.

Some highly desirable crops make good seeds, but planting those seeds would not result in more of the same. Apples and apple trees are a prime example. One of my favorite eating apples is Honeycrisp. If I ate a Honeycrisp apple and planted its seeds, they would likely germinate. If I let them grow for 20 years, the trees would likely make apples. But

The Gyroscope of Life

they would not be Honeycrisp apples. They would more likely be inedible. For multiple reasons, all apple varieties make seeds that are genetically different than the tree from which they come. Some of the reasons relate to stuff from Genetics 101. Others are a result of reproductive peculiarities that Mona built into apples.

The main reason Honeycrisp apples don't breed true – produce genetically identical offspring – is a roadblock that Mona set up. Most apple trees cannot be fertilized by their own pollen. Honeycrisp flowers are unable to fertilize themselves or any other flower on the same tree. ("Fertilized" in this case has to do with sexual fertilization, i.e., birds and bees stuff, not adding fertilizer.) Bees will often pollinate flowers on the same tree, but the pollen fails to grow, or fruit formation is blocked. For reasons to be explained shortly, all Honeycrisp trees are genetically identical down to the last gene, to include the genes that control Mona's anti-incest policy. So pollen from another Honeycrisp tree would be rejected also. Apple growers know this, of course. To get a good crop of Honeycrisp apples, they must put non-Honeycrisp varieties in the orchard. Crabapple trees are now often used as the pollen producer in commercial orchards. My delicious Honeycrisp apple may have seeds that are genetically half crab apple. If so, its seeds would produce nothing close to Honeycrisp fruit.

In many agricultural schools, apple breeders teach Genetics 101. They understand completely the futility of trying to produce true-breeding apple varieties. Instead, orchardists who grow apples, plums, apricots, peaches, and many other fruits use vegetative propagation. They cut off twigs or buds from a Honeycrisp tree or a Valencia orange and graft them onto another apple or orange tree. If you're wondering, the

Species, Evolution, and Domestication

original Honeycrisp was the product of a breeding program, but the only way to keep the product intact genetically has been vegetative propagation. Every Honeycrisp tree, wherever it may be grown, is a clone – an exact genetic copy – of that original Honeycrisp tree.

Breeders of corn have a similar problem. Per Genetics 101, it is essentially impossible to breed a consistently productive field corn line whose seeds can be replanted with equal productivity. Hybrid corn breeding, developed in the 1930s, moved corn production ahead by leaps and bounds. This revolution started with inbreeding corn lines. Inbred lines are less productive, but they can be reproduced time after time. Producers of hybrid corn seeds plant two inbred lines in the same field and then harvest the hybridized seeds resulting from the cross-pollination. In that way, they produce time after time a highly productive hybrid line of field corn. Such commercial lines exhibit what is often called "hybrid vigor".

In all of these cases – seedless fruits, genetically irreproducible apples, and hybrid corn – the basic strategy is to lock in the desirable genetic makeup and block further genetic change. That might seem to run counter to the pattern described earlier, where we have taken advantage of genetic mutability, but these are Goldilocks cases where the porridge temperature, chair height, and bed firmness are just right, causing us to want to lock in those traits, even if we have to go out of our way to make it happen.

In short, species are not fixed in time. They are mutable. That makes them excellent – and inevitable – candidates for evolution. It also makes them easy victims of human manipulation. Looking at them less sympathetically and more selfishly, we can say that the mutability of species has worked

The Gyroscope of Life

wonderfully to our advantage in providing grist for the mills of plant and animal breeders and for those first prehistoric agriculturalists. In the next chapter, we will look at natural species as actors on Mona's stage: the environment and its ecosystems. We will also look at the considerably altered stage upon which domesticated animals and plants play. The goal is to examine processes that work wonderfully in nature and see if or how they might be incorporated into agricultural settings.

10

Ecology, Ecosystems, and Agroecosystems

We travel together, passengers on a little spaceship, dependent on its vulnerable reserves of air and soil; all committed, for our safety, to its security and peace; preserved from annihilation only by the care, the work, and the love we give our fragile craft.

Adlai Stevenson II (1900-1965)

House Rules and *Homo sapiens*

Chapter 5 laid out the three House Rules Mona imposes on all her children – three requirements for remaining an actor on life's stage. Violate any Rule, and the hook comes out. What can we say about our own species and how we are faring on the stage? We clearly are doing well with regards to HR#3; we're making oodles of babies and thereby passing along our DNA. That same piece of evidence establishes that we are collectively handling HR#2 nicely; we are keeping the species' entropy low. But what about HR#1 – fitting our environment – the ecological imperative? That is not so easy to assess. The ecology of modern humans is unlike that of other species. We have a distinctly different relationship with our environment. This chapter will describe how different that relationship is and then look again at HR#1.

The evolution of ecology and environmentalism

I have sometimes passed myself off as an ecologist. I've

The Gyroscope of Life

even taught a couple of courses with ecology in their names. But I am not trained in the field. In my college years, ecology wasn't widely recognized – not even among biologists. Ecological concepts have floated around since Aristotle, but those ideas began to be organized into a distinct biological field only in the second half of the 19th century. The word ecology was coined about 1870, and it became an academic subject by the dawn of the 20th century, but it caught on slowly. The first American textbook on ecology was written in 1953 by the Odum brothers at the University of Georgia.

Mainline biologists may have had some initial resistance to this new branch of biology. They perhaps considered ecology to be something left over from the days of natural history. But, in the second half of the 20th century, ecology evolved to become coequal with other biological disciplines. The Manhattan Project, which spawned the first atomic bombs and then morphed into today's US Department of Energy, played a significant role in raising the profile of ecology in America. Scientists were needed to document the biological disturbances caused by nuclear blasts. Beginning in the 1950s, lots of federal funding went into ecological studies, and that helped to popularize and populate the field.

The histories of environmental and conservation movements are closely intertwined with ecology. From early times in America, the land and its resources were seen as commodities to be consumed. That use-it-up, wear-it-out ethic persisted till well after the west was fully settled. But, by the mid-20th century, the consequences of such an ethic began forcing itself onto the cultural consciousness, not just in the US, but globally. The Dust Bowl era of the 1930s showed how serious human impacts on the environment could be. In 1952, the 5-day-long Great Smog of London was blamed for

Ecology, Ecosystems, and Agroecosystems

4,000 deaths, and another 6,000 subsequently died because their health had been compromised by the toxic air. In 1969, the highly polluted Cuyahoga River in northeastern Ohio caught fire again (reputedly for the 13th time in 100 years). Also in 1969 came the Santa Barbara oil spill, when an offshore oil well blew out. The sludge-covered California beaches and dead birds and sea lions were lead items in the national news for several days. These and other ugly stories showed how badly modern societies sucked at environmental stewardship, and the growing field of ecology offered even more examples of the negative impacts of humans on living things.

Aldo Leopold helped raise America's environmental consciousness. He is known especially for *Sand County Almanac*, which was published posthumously in 1949, with over 2 million copies printed. Leopold was a wildlife biologist and ecologist in a time when ecology was still coming out of the shadows. He wrote passionately about the need for a "land ethic" – a realignment of the relationship between people and the environment. One of the essays in the book is entitled "Thinking Like a Mountain". Reading it an umpteenth time can still bring tears to my eyes and a chill down my spine. Leopold has been described as a nature writer. My reading of *Sand County Almanac* left me feeling he was a poet who wrote in paragraph form. One of his quotes is: "A thing is right when it tends to preserve the integrity, stability, and beauty of the biotic community. It is wrong when it tends otherwise." We will come back to that notion (and to "Thinking Like a Mountain") in another few pages.

Then there was Rachel Carson. Her *Silent Spring* (over 6 million copies sold in the US) was published in 1962, two weeks after I went off to college. Carson was a marine biologist and an ardent conservationist. Her book, which was a

The Gyroscope of Life

catalog of environmental transgressions, grabbed society's attention. Industry almost unanimously attacked the book's message, but Carson had amassed too much data to be dismissed. She died in early 1964, while the debate still raged, but the impetus for change that she created continued to grow. Her book galvanized federal environmental legislation – most especially the National Environmental Protection Act of 1969, and that momentum carried into succeeding administrations. Despite foot dragging by industry and occasional reversals of environmental standards, the electorate still seems to favor clean air, clean water, and springs when birds sing.

So, over my lifetime, I have watched the growth of ecology, environmentalism, and conservationism. They were not widely recognized when I was born. They began to find acceptance, even legitimacy, in my school years. They are now household words. More recently however, environmentalism and conservationism have become pejorative terms in some circles. Various forces have worked to restore the ethic that resources are here to be consumed. Indeed, it is a theological issue for some; God commanded newly created mankind to "Be fruitful, and multiply, and replenish the earth, and subdue it: and have dominion over the fish of the sea, and over the fowl of the air, and over every living thing" (Genesis 1:28). Tension between subduing, having dominion, and conserving is inevitable – especially as we become increasingly fruitful and multiplied.

In spite of its general acceptance among biologists now, ecology still has something of an identity problem. It remains a rather ill-defined field. Ecologists are much like the blind men sizing up an elephant. Many have their own take on what the field looks like, and it can be difficult to find

Ecology, Ecosystems, and Agroecosystems

overlap between their views. For purposes of this chapter, I will cast our lot with those who define ecology as the study of communities of organisms and their interactions with one another and their environment. This chapter deals with ecology, but we will touch only the hem of that garment. I will draw attention to four ecological concepts: ecosystem, sustainability, niche, and carrying capacity. They will then be used to assess how well humans are dealing with HR#1, while showing how different we are as ecological entities.

Ecosystems, fundamental units of ecology

When I hung my professional shingle in 1977, I taped on my door a lovely water-color poster. It featured a great blue heron at a pond's edge with plants, insects, and other animals in the scene. The single word on the bottom was "Ecosystem". I'm embarrassed to admit it, but the word was new to me then. I remember thinking it didn't seem to have been created by a biologist; it didn't sound Greek enough. Hell, I could tell almost intuitively what it meant. As it turned out, the word had been kicking around in ecological circles for about 40 years. But, then, I'd never had a course in ecology.

Some ecologists approach their studies as reductionists. They describe the field as one made of successive levels or layers of complexity, with each higher level having more parts. We introduced that sort of parsing when we discussed the philosophy of reductionism. In a reductionist's dissection of ecology (Table 10.1), an organism – a single living thing – constitutes the lowest level of consideration. All of the organisms belonging to a particular species that occur at a location of ecological interest make up a population, the next layer of complexity. All of the populations of all the different

The Gyroscope of Life

species found at that location – plants, animals, fungi, bacteria, etc. – equal a community. When we add to our consideration all the abiotic – non-living – components within that location, we have an ecosystem. After ecosystem comes a level of complexity often called biome. These are large geographic areas that support similar ecosystems, such as the millions of acres dominated by grasses on North America's prairies, South America's pampas, and Africa's savannas. If we shove all biomes into a final, highest level of complexity or organization, we get the ecological equivalent of the Theory of Everything. This globe-encompassing collage of life has been dubbed the biosphere, or ecosphere.

Table 10.1. A rather reductionist look at ecology. Each succeeding layer encompasses the components of the previous one. Ecosystem is the primary focus in this chapter. Some ecologists look at life holistically, as a planetary phenomenon.

Level/Rank	Description
Biosphere/Ecosphere ↑	All life on Earth plus the abiotic environment
Biome/Life zone ↑	A geographic region with similar ecosystems
Ecosystem ↑	A community plus its nonliving environment
Community ↑	All the populations of all species at one place
Population ↑	All the organisms of one species at one place
Individual organism	A single animal, plant, fungus, etc.

Ecology, Ecosystems, and Agroecosystems

Ecosystem is a somewhat elusive concept, but it can be readily intuited by anyone who has spent time outdoors in natural areas. An ecosystem is, in essence, the living room for – the space occupied by – a community of interacting plants, animals, fungi, and microbes. Wetlands are a common type of ecosystem. The community of organisms that live and interact in a wetland is distinctive; all its members are adapted to the conditions in their soggy living room. In fact, many of the species there occur only in wetlands. As a minimum, all the organisms there are adapted to their wet environment, and they interact with one another. Some interact in obvious ways and others in much subtler ways. The same can be said when we look at the organisms in forest, grassland, marine, freshwater, etc. ecosystems. Each ecosystem is unique physically, biologically, and ecologically. The ecosystem is the environment within which organisms must fit a la HR#1. In fact, HR#1 can be restated as "Fit your ecosystem."

Reviving the theatrical metaphor, an ecosystem is a set on the stage where life is being played out, and where Mona is the no-nonsense stage manager. Our theatrical metaphor is strained, however, when we look more closely at the relationship between the players and the set in an ecosystem. In theater, the stage is set by behind-the-scenes persons, and then the players come out and perform. But, in an ecosystem, the players (the biological community) serve a major role in creating and maintaining the set. The living things are functional parts of the set, of the ecosystem. For example, dam-building beavers alter stream flows such that wetlands are formed. The animals and plants found in these beaver-engineered ecosystems will fit (be adapted to) the environment created by that hydrology. They all contribute to reshaping or maintaining the set as the play goes on.

The Gyroscope of Life

Pacific salmon provide a stunning example of how a species can have major impacts on the life-sustaining properties of an ecosystem. These fish hatch in freshwater streams of the Pacific Northwest and then spend most of their lives in the ocean. When they are ready to spawn, they head home to where they hatched. They run upstream, spawn, and then they die. That final act, dying, delivers nutrients and chemical energy back into headwaters and surrounding ecosystems. (On their way home to spawn and die, salmon also provide nutrients and energy to many predators and carrion feeders: bears, eagles, osprey, etc.) Native Americans used fish to fertilize their crops. Mona has been doing the same thing in and around salmon streams for tens of thousands – maybe hundreds of thousands – of years. The plants and animals found there are a testament to the salmon's contributions to what would otherwise be a less fertile and less diverse environment.

In short, living things are active participants in the development and maintenance of their ecosystems. The combination of living and nonliving components that make up an ecosystem provide gyroscope-like stability and perpetuity, or sustainability. We must explore the notion of sustainability before coming back to consider some dramas playing out on ecosystem stages and before discussing how different an ecological actor is *Homo sapiens*.

Sustainability and the ecological Gyroscope of Life

Ecosystems are not forever, but they can persist in a stable form for many millennia. Over long periods of time, as we humans see time, ecosystems can be remarkably unchanging.

Ecology, Ecosystems, and Agroecosystems

Turnover occurs. Organisms die and are replaced, but the mix of organisms today can be just as it was 10,000 or even 50,000 years ago. A figurative Gyroscope of Life maintains stability at the ecosystem level. The Gyroscope's flywheel is the community, the living components of the ecosystem. The energy source that keeps the wheel spinning is, in most cases, the Sun. As long as their Gyroscopes operate smoothly, healthy ecosystems are sustainable – literally, able to sustain themselves quasi-permanently.

The only "inputs" needed to keep most ecosystems operating perpetually – to keep their Gyroscopes spinning smoothly – are air, water, sunlight, soil, nutrients, and a healthy community of organisms. Inputs is in quotes because those resources are generally considered inherent to the ecosystem and not inputs per se. With those ingredients, an ecosystem can make and maintain lots of living stuff – the communal flywheel – and do so indefinitely. Indeed, natural ecosystems must operate by this self-reliant, non-exhaustive, no-external-inputs ethos; they have no alternative. There is no resupply for ingredients that might become exhausted. Salmon streams and surrounding ecosystems might seem an exception, but the salmon that return and die are an integral part of those ecosystems.

In tropical rainforests, some of Earth's most productive ecosystems, the mineral nutrients needed by plants and animals are generally in short supply. At any given moment, most of the life-essential minerals are tied up in living matter. High temperatures, abundant rainfall, and infertile soils put a premium on plants being able to recapture and recycle nutrients as soon as a leaf, tree, or animal falls to the forest floor and starts to decompose. The nutrient cycling system in a rainforest is described as being tight, or even a closed loop.

The Gyroscope of Life

Networks of plant roots develop at the soil surface and can absorb nutrients as soon as they become available during the breakdown of dead matter. That keeps the ecosystem non-exhaustive, sustainable, able to operate over the long haul. That helps maintain the stage setting. It keeps the room livable.

The grasslands on the North American Great Plains were remarkable for their huge herds of buffalo, deer, and antelope. Grasses, which had all the air, sunlight, water, and nutrients they needed, grew abundantly and turned the prairies green. The grazing animals ate those grasses and returned the nutrients in short order as manure or over the longer haul as dead carcasses. But, if all those animals had gone elsewhere to crap and die (yeah, unlikely, given the scale), those nutrients would have been lost from the ecosystem. As a result, grass production would have fallen, animals would have starved, and the ecosystem's Gyroscope would have been destabilized. The very fact that rain forests, prairies, and all other ecosystems can persist unchanged for thousands of years testifies to their non-exhaustive, sustainable *modus operandi*.

In short, natural ecosystems – those communities of distinctive organisms living together in a particular place – are notable for their sustainability. They persist in their setting and they do so without reducing the likelihood that they can continue to do so indefinitely. They are, in a sense, perpetual motion machines. As long as the sun shines and the rain falls, they just keep going and going and going.

Niche: Assigned roles in ecosystems

Niche, properly pronounced "nitch" or "neesh", is a rich ecological concept whose meaning is often clouded by its

Ecology, Ecosystems, and Agroecosystems

casual misuse. It is often misapplied as being essentially synonymous with habitat, the type of environment for which an organism is well-adapted. It is also misused to describe the particular space or geography that an organism occupies, its territory. But niche is about lifestyle, about how an organism earns its living and how it fits into its community and ecosystem, using fit in the same sense that Mona uses in HR#1. An organism's niche is its unique functional role in its ecosystem. Each species – each population – in an ecosystem has a different job. Each plays a unique part in keeping the flywheel balanced and spinning smoothly. All fit in accordance with HR#1 and contribute to the overall sustainability of the ecosystem.

Tennyson observed that nature is "red in tooth and claw", and Darwin characterized competition between living things as a driving force for evolution. Some bring those two classic observations together and think of predators and their prey as being in competition. They are not. Each is playing a completely different role in their ecosystems. Each fills a distinct niche. Predators are more likely to compete with one another for the prey they seek, and prey animals are more likely to compete with one another for the food that they eat. In fact, predators can help the prey by "thinning the herd" and reducing prey's competition for food. Smoothly running ecosystems typically rely on some animals filling niches where they eat vegetation, while other animals fill niches in which their role is to eat the vegetarians. Predation keeps healthy ecosystems balanced and stable. The ecological importance of predator niches in maintaining ecosystem sustainability – in keeping the Gyroscope spinning smoothly – has been amply, yet inadvertently, demonstrated in Yellowstone National Park.

The Gyroscope of Life

In the early 20th century, a federally funded program used poisons and bounty hunting to eliminate gray wolves from the so-called Greater Yellowstone Ecosystem (GYE). (The GYE is a collection of otherwise distinct wetland, grassland, forest, and freshwater ecosystems. Yellowstone National Park makes up about 10% of the GYE.) Wolves are smart, social predators that, in a pack, can bring down a buffalo, but they prefer elk. Unfortunately for them and ranchers, they do not discriminate between domesticated and wild prey. The last wolf in the GYE was killed in 1926.

As it turned out, wolves were vital to the health and sustainability of the GYE. When they disappeared, the Park's ecology was radically altered. In their absence, and presumably because of their absence, wintertime elk herds in the Park grew from around 5,000 in the first half of the 20th century to around 19,000 in the 1980s to 1990s. This elk boom proved to be unsustainable. Many stands of aspen trees in the Park became stunted, as starving elk herds were forced to browse on aspen in the winter. At the same time, shrubs growing around Yellowstone streams were being browsed into oblivion. Beaver populations declined, because the now-missing shrubs had provided beavers food as well as material to build their dams and lodges. Without the beavers' dam building, wetlands began to dry up, and other wetland species disappeared. In short, plants and animals that shared the GYE with elk experienced serious problems because the wolf population disappeared and the elk population consequently got out of balance. It was as though a piece of the flywheel that was the wolf's niche flew off and a big lump arose on the portion of the flywheel that was the elk's niche. The Gyroscope got wobbly.

(Aldo Leopold's epic – but short – poetic essay "Thinking

Ecology, Ecosystems, and Agroecosystems

Like a Mountain" is about the relationship between wolves, their prey, and the health of ecosystems. He contritely admits to, in his youth, being party to the notion that the only good wolf is a dead wolf. But he goes on to note how short-sighted, even ignorant that notion was. As he began to "think like a mountain", he understood that wolves were vital to the health of not just the mountain and its ecosystems but to the wolves' prey as well.)

Forty-one wolves were controversially reintroduced into Yellowstone Park in the mid-1990s. Over the next 15 years, wintertime elk populations declined to about 5,000 again. The wolves were dining well, and their numbers grew. And other good things were happening. The stunted stands of aspen trees began to recover. Woody stream-side vegetation regrew. (Some ecologists have hypothesized that, with the return of wolves, elk changed their feeding behavior and stopped browsing around streams flowing through open, grassy meadows. A pack of wolves can race across that open space more quickly than a herd of elk.) As the stream-side shrubs regrew, beavers came back. The beavers' dam building re-established wetlands that supported the return of unique wetland organisms. Moose returned to their old haunts. Fish populations increased. A balance and diversity were reestablished – balance and diversity that had been lost at least partially by obliterating wolves from their niche. The wolves' niche had remained. The job had remained, but, for about 70 years, wolves were not there to take care of business. With their return, the ecological Gyroscope re-stabilized.

The Yellowstone wolf-elk case is perhaps not as open-and-shut as the last few paragraphs might suggest. We see definite effects (changes in elk herd size and vegetation), but we cannot be sure of all the causes. For example, the elk irrup-

The Gyroscope of Life

tion didn't occur until about 40 years after wolves had been eliminated. So, the elk boom wasn't due solely to the absence of wolves. (Grizzly bear numbers in the GYE dropped to the point that they were declared an endangered species in 1976, and bears are major elk predators. That may account for some of the delay in the elk population boom relative to the disappearance of wolves.) Some point out that the GYE was just coming out of a years-long drought about the same time the wolves were reintroduced, and that could account for at least some of the vegetation recovery. Still, many ecologists tell this story as a cautionary tale, one that shows unintended consequences of "managing" ecosystems without understanding how ecosystems work or how unraveled the garment may become when only one thread, or niche, is pulled out.

Another destroyed-niche story is unfolding in Yellowstone. In this case, it is a prey's niche that has been obliterated. The problem was created by the introduction of an exotic species. Lake trout are great fun to catch; so much so that someone – not a federal official – is suspected of introducing them into Yellowstone Lake in the 1980s. As it turns out, lake trout feed voraciously on cutthroat trout, the beautiful native trout that have been a linchpin in Yellowstone ecology for thousands of years. During spawn, cutthroat from Yellowstone Lake migrate into its tributaries and provide a major source of food for grizzly bears, eagles, pelicans, otters, osprey, etc. With the introduction of lake trout and the consequent loss of about 95% of the cutthroat population, the niche, or role, served by the cutthroat was essentially vacated. Eagles, which would normally feed on cutthroat, have been observed to now prey on young trumpeter swans, an endangered species, and on other water fowl. With the loss

Ecology, Ecosystems, and Agroecosystems

of cutthroat from their menu, grizzlies are now more likely to prey on elk calves. National Park Service personnel are working to net and destroy lake trout in the 136-square-mile lake, and they appear to be making some headway.

Mona builds a balance into well-functioning ecosystems such that predator and prey numbers are proportionate. Wolf numbers in the Park rose to about 150 after they were re-introduced, but they now number nearer 100; presumably because 100 wolves to 5,000 elk provides a better predator-to-prey balance. Unfortunately, there is no lake-trout-to-cutthroat-trout balance for Yellowstone Lake – at least not one that also feeds the rest of the cutthroat's predators. Hence the $2-million-per-year effort to reduce the population of lake trout and restore cutthroat to their niche.

Niche: Minimizing competition and achieving détente

Living things compete for resources. Elk compete with elk and with other species for browse, and trout compete with one another for insects and minnows. Competition is a theme for plants, too. Green plants everywhere compete for space and the sunlight that comes with it. Water is a crucial resource for every living thing. In some environments, a drop obtained by one organism could mean that another organism will not survive. Competition for resources is inevitable as plants and animals strive to fit their environment, keep their entropy low, and make more of themselves.

But niche generally allows the various species in a healthy community to interact in ways that are more nearly cooperative than a life-or-death competition. Rather than making real estate something that must be fought over, niche pro-

The Gyroscope of Life

vides strategies for achieving détente between species that share the same living room. In a healthy ecosystem, many different plants and animals can live check by jowl, and, in many cases, they would not live so well were they not living together. In filling different niches within an ecosystem, organisms reduce competition and achieve some measure of peaceful coexistence.

Plants provide numerous examples of competition-avoiding niches. By employing different growing seasons, or times of greatest activity, plants can live side by side and minimize the competition for resources such as light and water. Some plants living on the forest floor remain green in the winter and can take full advantage of the sunlight passing through the bare branches above them. Some plants appear in early spring and can complete their crucial tasks of flowering and making seeds before trees' leaves appear. Bloodroot, trilliums, and spring beauty are some of my favorite Appalachian early-spring bloomers, but they are pretty much gone by the time leaves appear above them. Skunk cabbage blooms in January and February, so early that shade from deciduous trees is never a problem, but the flowers can sometimes end up under snow. This relative of calla lily and jack-in-the-pulpit has a good strategy for regaining its place in the sun, though. It is one of a few plants that can raise their temperature above the ambient. Skunk cabbage can melt a hole in the snow around it. In all of these examples, small plants have found and fit into niches that allow them access to light and water, while living with much bigger plants.

Niche also allows plants and animals to live communally by taking advantage of the fact that environmental conditions vary within an ecosystem. Spots that are only a few feet or inches apart may be more or less sunny, be drier or wet-

Ecology, Ecosystems, and Agroecosystems

ter, be more acid or more alkaline, have deeper or shallower soils, and possess more or less of some key nutrient. Some plants and animals may find those differences much to their liking. Some will fit into unique niches by being adapted to more acidic spots, to drier spots, to shadier spots, etc. It's the Goldilocks thing with Mona. She says in essence, "I don't care where you sit, just make sure you fit the spot you pick".

In the northern hemisphere, north-facing slopes tend to be cooler, shadier, and wetter than those facing south. In the Appalachians, one can readily see the niche effect of northern versus southern exposure. North-facing slopes are often covered with pine and rhododendron, while the south-facing slopes will be dominated by deciduous hardwoods. The effect is most striking after the leaves fall. In all ecosystems, plants and animals that are better adapted to conditions that are wetter or dryer, cooler or warmer, lower or higher pH, etc. can find a niche for themselves and increase overall diversity, while reducing competition. Niche is why mosses favor the north side of trees in northern climes and the south side in the southern hemisphere.

Lest we paint too rosy a picture for détente and peaceful coexistence, let's note that some organisms are openly hostile and competitive with other members of their community. Within the animal kingdom, many insects produce toxins that make them inedible. Many plants have also developed chemical defenses against herbivores. In fact, insects and plants have been in an evolutionary arms race with each other for hundreds of millions of years, with each side continually shifting strategies. Insects evolved from just chewing on plants to sucking juices out of them and then to burrowing into them. Some insects can also cause plants to produce swellings called galls. These galls, which can be the size of a

The Gyroscope of Life

golf or tennis ball, serve to protect and nurture the insects' larval offspring. "Oak apples" are produced by oak trees when gall wasps lay eggs on their leaves. Oak apples were a major source of writing ink from Roman times into the early 20th century. Iron gall ink from oak apples is still being made commercially for fountain pen purists. But I digress…

Plants were evolving, too, in the evolutionary arms race with insects. After they moved onto land, plants evolved and weaned themselves of the need for iodine. That was a good move, because iodine is scarce on land. However, all animals, to include insects, have continued to require iodine as an essential nutrient. Some plants have learned to use that iodine dependence against insects, producing chemicals that block iodine uptake. Those chemicals are not self-toxic, since plants don't need iodine, but it is a different story for any insect who might chew, suck, or burrow into those plants.

Some plants make cyanide-producing chemicals and other poisons, causing the plant to be toxic or distasteful to would-be diners, whether insects or higher animals. Many of those additions to the plants' arsenal have ended up in our medicine cabinets and/or have found other commercial, culinary, or recreational uses. *Digitalis* is the Latin name of plants commonly called foxglove, and it is the name of a chemical made by foxglove plants. When administered in proper dosages, digitalis is a valuable medicine for treating some human heart conditions, but at higher dosages, it can be toxic. Insects and animals can get those toxic dosages when they feed on foxglove. Other chemicals that plants use to discourage or defeat insects and other herbivores include many that we have adopted for our own purposes. Those include nicotine, caffeine, morphine, quinine, citronella, menthol, and the cannabinoids such as THC. If THC doesn't ring a bell,

Ecology, Ecosystems, and Agroecosystems

perhaps you've never used marijuana – or perhaps you've used too much.

And some plants produce chemicals that are toxic to other plants. Their leaves and/or roots produce substances that interfere with the growth of other species. Such plants are common in desert ecosystems. A plant using this tactic keeps other plants from growing in its territory. Aerial photographs give them away – showing a plant in the middle of an otherwise clear circle. What might seem an act of hostility or aggression – blocking the growth of other plants – could be seen autobiologically as just defensive. The xenophobic plant, which needs X amount of water to survive, is staking out its territory. That circle we see from the air perhaps receives X amount of rain in a typical year – not enough for another plant. The plant's chemical cold shoulder "encourages" potential competitors to live elsewhere.

In Appalachia, a very common species that carries on this kind of chemical warfare with its botanical neighbors is black walnut. Black walnut trees produce a substance that stunts the growth of plants in their vicinity. It works well for the walnut. Not so well for other species (or for an unnamed gardener whose neighbor's walnut tree hung over his garden).

In any event, these examples of semi- or pseudo-hostility are more the exception than the rule. Most living things appear quite content to fill their unique niche and to be neighborly. They seem to recognize – or more accurately, to have evolved to interact with – their cohabitants as integral parts of their ecosystem's Gyroscope and essential to the commonwealth.

One more mega-concept: Carrying capacity

The Gyroscope of Life

Carrying capacity is an ecological concept tied to ecosystem, sustainability, niche, and HR#1. The Yellowstone wolf-elk story grimly illustrates the concept of carrying capacity. After wolves were exterminated from their niche in Yellowstone, the Park's wintering elk herd almost quadrupled. The consequences were bad for the elk and for the ecosystem. That many hungry elk threw the Gyroscope's flywheel out of balance. Elk were forced to expand their menu and began to eat things not usually part of their diet, but they still suffered from starvation and greater calf mortality. The Park's ecosystems simply could not support 19,000 elk. Its carrying capacity for elk was exceeded.

Carrying capacity suggests an ecosystem can sustainably, or perpetually, support only so many individuals of a species. A sustainable population is one the ecosystem can provide sufficient resources for year after year after year. When carrying capacity is exceeded and an ecosystem is overtaxed, the population "in violation" will find itself competing with itself for resources – often resulting in mass starvation. Having too many weakened individuals crowded together may also allow an otherwise minor disease to become epidemic. In either case, exceeding that sustainable number – exceeding the ecosystem's carrying capacity – has bad consequences for the offending species. But a price may be paid by the entire ecosystem as well. Overpopulation by a single species can lead to an overall loss of ecosystem balance and sustainability. The ecological Gyroscope becomes unbalanced and begins to wobble.

Exceeding carrying capacity is, in essence, a violation of HR#1. Otherwise well-adapted organisms no longer fit their environment if they over-produce to the point that they overtax ecosystem resources. If they reproduce too exuber-

Ecology, Ecosystems, and Agroecosystems

antly, Mona will correct them. The correction may not be extinction, but, as a minimum, something unpleasant – starvation, disease, reduced fertility, etc. – will happen to the offenders and cause population numbers to drop back into a range that can be sustained by the ecosystem – within its carrying capacity.

Here's where we've been heading: Does carrying capacity apply to *Homo sapiens*?

So, now comes the existential question we've been building toward: do we humans have to obey HR#1 and stay within our ecosystem's carrying capacity? Are we subject to the same consequences that those elk in Yellowstone Park were? An accurate, but uninformative, answer would be "maybe a few are". A perhaps technically correct, but snarky, answer would be "the question is irrelevant, since most humans do not belong to an ecosystem". While these answers are true, we should take no comfort in either of them.

Carrying capacity applies to *Homo sapiens* whenever humans are fully integrated into an ecosystem. Ancient hunter-gatherers fit, a la Mona's HR#1, into various forest, grassland, wetland, tundra, etc. ecosystems – just filling their niche and being part of the Gyroscope. Today, some humans may still approach that truly at-one-with-nature life style. Inuit and Yupik peoples in the Arctic, hunter-gatherers in Africa, Amazonia, and Oceania, nomadic peoples in parts of Africa and Asia, and subsistence farmers in many places have lived such lives in the near past, and some still may. They can fill a niche within their ecosystems and be just as subject to HR#1 as every other species. If they exceed their ecosystems' carrying capacity, there can be consequences.

The Gyroscope of Life

But a bit of autobiologic suggests the great, great majority of modern humans are not integrated into any ecosystems. We impose ourselves on them, but we do not fill a positive ecological role in most of the places we choose to live. We are freelancers, living niche-less lives, outside the bounds that Mona envisioned for her children. Most of us are essentially ecological interlopers – exotic, invasive species and not functional parts of an ecosystem. Our ingenuity and technology let us live in Death Valley, Antarctica, and many other places for which we are entirely unfit biologically. Modern medicine has extended our lifetimes and provided us relief from scourges that were perhaps just one of Mona's ways of keeping ecosystem Gyroscopes spinning smoothly.

In short, we are essentially living niche-free and ecosystem-less lives, and our numbers continue to grow. Have we escaped Mona's rule? Perhaps HR#1 and the notion of carrying capacity no longer apply to us? (Was that what those Yellowstone elk were thinking as their population quadrupled?) If something comparable to a carrying capacity exists for modern, mobile, niche- and ecosystem-free humans, it must occur somewhere above the level of ecosystem (Table 10.1). The only likely level is the biosphere, but can we justify making such a shift in perspective? Is the planet an ecological entity with checks and balances and self-regulating feedback controls? Does Earth have a Gyroscope of Life? If so, Mona could have something in store for a species that disrupts its balance.

Over the last 2-plus billion years, Earth's global environment has been shaped dramatically by life – by myriad ecosystems spreading across the globe. Atmospheric oxygen levels, greenhouse gases, the ozone layer, ocean salinity, the hydrologic cycle, global temperature, and other global-level

Ecology, Ecosystems, and Agroecosystems

phenomena have been heavily influenced – some would say regulated – by living forms. There is good evidence, for example, that the level of carbon dioxide in the atmosphere has long been regulated by living things. For the last 0.8 billion years, atmospheric carbon dioxide levels have jiggled between 200 and 280 parts per million (ppm), with the lower values occurring during ice ages. Throughout that time, the concentration of carbon dioxide in the atmosphere reflected the balance between photosynthetic uptake and respiratory release of the gas by the collective of all living things.

The ability of living things to regulate planet-level processes is also seen in Earth's oceans. Geologists and hydrologists tell us that, over the last half billion years, the fraction of seawater that is salt has remained pretty steady at around 3.5% – a life-friendly amount for organisms that evolved in and are adapted to that environment. But we might have expected salinity to be increasing because additional salts are continually flowing from land to sea. In some landlocked bodies of water, e.g., the Dead Sea and the Great Salt Lake, salt now constitutes one-third of their water's content. Much of the explanation for the stable and relatively low levels of salt in Earth's open seas appears to lie in the ability of certain ocean-living bacteria to convert soluble salts into insoluble minerals, lowering salinity. The bacteria's role is analogous to how our kidney's control salt levels in our blood. It demonstrates powerfully the ability of life to regulate conditions at a global scale.

Because the collective of life – the biosphere – generates and/or regulates many life-sustaining properties of the planet, it may not be too much of a stretch to view the biosphere as one huge, self-regulating, Gyroscope-stabilized meta-ecosystem, or even a single superorganism. The no-

The Gyroscope of Life

tion of Earth as a self-regulating superorganism – the Gaia Hypothesis – was suggested in 1979 by James Lovelock, an English polymath. Whether meta-ecosystem or superorganism, some version of ecology would appear to be going on in the biosphere. That opens the door to the notion of reapplying the concept of a biospheric carrying capacity for a species that recently escaped such consideration at a lower level.

Accordingly, a biospheric carrying capacity for *Homo sapiens* would be the number of us that could live on Earth sustainably – century after century, millennium after millennium – without reducing the planet's ability to support another generation. Put another way, Earth's human carrying capacity is the upper limit of how many people can live on Earth without doing harm to the planet such that it can no longer support that many. For modern humans, HR#1 will read "Fit your ecosphere." Along with that goes the admonition, "Do not exceed the planet's carrying capacity."

We know that we have the potential to do harm to the planet's ability to self-regulate and to support – "carry" – us. Since the Industrial Revolution, our species has overwhelmed the biologically regulated level of carbon dioxide in the atmosphere. The current level is about 415 ppm, and that number will be higher by the time you read this. Rather than contributing to the stability of this key indicator, we are pushing it into new territory. And that is just one indication of the impact we humans are having on a global scale. Species extinctions, driven by over-harvesting and human-driven habitat loss, are occurring at increasing rates. (We may not know how a species is important, and maybe its ecosystem can get along without it, but we are playing a game of ecological Russian roulette with each click that signals another species' extinction.) Oceanic garbage patches now cover

Ecology, Ecosystems, and Agroecosystems

huge areas, with one in the South Pacific being larger than Mexico. Clearing and burning of Amazonian rainforests are obliterating big chunks of what has been described as the "lungs of the planet". We humans are clearly having ecological impacts – mostly negative – on ecosystems and the ecosphere. And that will have implications for how many of us the planet can support or carry.

How many is too many?

Thomas Malthus was a 19th century economist, not an ecologist. Malthus never heard of carrying capacity or the biosphere, but he had much to say about human overpopulation. Interestingly, he was a crepehanger who saw the crepe as a silver lining. He posited that the capacity to produce food grows only arithmetically, while population grows geometrically, or exponentially. For him, human overpopulation was inevitable and inevitably would lead to disease, starvation, and/or war. Those "positive checks" (his term) would bring populations back to a size whose needs could be met. He blithely observed, "The scourges of pestilence, famine, wars, and earthquakes have come to be regarded as a blessing to overcrowded nations, since they serve to prune away the luxuriant growth of the human race." Modern economists have declared Malthus' ideas unnecessarily pessimistic or irrelevant in a technological age. However, his basic thesis is sadly right on the mark today in places where birth rates are high and regional food production has fallen hopelessly behind demand. In such places, the consequences sadly echo Malthus' predictions of famine, pestilence, and war. From an ecological perspective, these are situations where humans have exceeded their local environment's carrying capacity –

The Gyroscope of Life

and Mona is enforcing HR#1.

Some argue that such problems are localized, and that Earth's – the biosphere's – food production capacity is sufficient to feed 7.7 billion of us. Even if they are right (and I will not concede that), food production capacity is not a proxy for carrying capacity. Humans put many more demands on the ecosphere than just to feed us. We need clean water and clean air, resources that we increasingly threaten. We need materials to make clothing and shelter. We extract resources to stoke our economic and technological furnaces, and, while doing it, we often cause ecological havoc, e.g., huge oil spills, mountains obliterated by coal mining, and vast areas denuded by "tailings" and "spoil", the leftovers from extracting and purifying ores.

Hence, the existential question comes again: Can we exceed Earth's carrying capacity? Some think that we can and that we already have. When there were 3.5 billion of us, Paul Ehrlich concluded Earth was already overpopulated, and he likened the situation to a burning fuse when he wrote *The Population Bomb* (1968). His arguments, which have been criticized by some – mostly economists, not ecologists – seem Malthusian, but with an ecologic spin. He and many more ecologists feel we are living on borrowed time – and at the expense of future generations, who will find it increasingly difficult to obtain sustenance and other needs in an overpopulated world. After 50 years as a biologist and an agronomist, during which time I've thought about this a lot, I have come to the same conclusion. We are in trouble. In a few more pages, I will put on my agronomist hat to explain why I think so.

Ecology, Ecosystems, and Agroecosystems

Are humans ecologically "malignant"?

My dealings with an aggressive form of cancer have raised my awareness of what can happen when a once-stable entity, a prostate gland, begins to grow willy-nilly and to disperse its cells into areas for which they are unsuited. Those spreading, rapidly multiplying cells are already disrupting components of my body's Gyroscope, especially my skeletal and lymphatic systems. In the not too distant future, perhaps before you read this, my Gyroscope will have gotten irredeemably out to kilter because of the cancer, and it will have flopped over. Could that be where the ecosphere is heading with our species? A case could be made that human beings, who once benignly filled some minor niches, have gained the ability to grow tumor-like and to metastasize throughout the ecosphere. Ecosystems are being altered. Species are being driven into extinction. Ecosystems' Gyroscopes are being destabilized.

Projecting that malignant metaphor into the future might suggest that the entire ecosphere will wither and die, that all life will disappear. I am persuaded that life is much too resilient for that fate. But it is not too much of a stretch to posit that we modern niche-free humans might disappear. That is because we "supermen", who can sidestep the restraints that Mona has placed on other living things, are spreading kryptonite all over the planet even as we enjoy our time here. We could be jiggering our own demise in the process, and the biosphere might be better for it. E.O. Wilson, an acclaimed entomologist and ecologist, has observed, "If all mankind were to disappear, the world would regenerate back to the rich state of equilibrium that existed 10,000 years ago. If insects were to vanish, the environment would collapse into chaos."

The Gyroscope of Life

The modern geologic epoch has been tentatively dubbed Anthropocene because of the enormous impact that human activity is having on Earth's ecosystems, geology, and climate; those impacts are seldom positive. That is not too many degrees off of the notion that humans are cancer-like. Mona would perhaps think so.

If we have gone wrong in Mona's view, how did it happen?

We can assume autobiologically that our hominid forbears filled distinct niches in the ecosystems to which they belonged. The role of each ancestral species was dictated by its physical and behavioral characteristics as well as by the other organisms in its ecosystems. Everyone fit, and the ecosystems were stable. Mona was happy. But at some point, pre-humans developed technologies that give them an edge over some other species in their communities. The oldest known hominid tools date back 3.3 million years. Those tool makers weren't *Homo sapiens* or even earlier versions of *Homo*, but they were heading toward humanity, and their evolving brains were capable of learning how to make tools from rocks. This started hominids down a technological path that has now widened into a superhighway, and the wider that path has gotten the more individuals have been able to tread it. We may have figured a way around Mona's HR#1.

The first humans – various species of the *Homo* genus – trod a narrow technological path. They likely inherited the making of stone tools, since that skill was present in pre-human ancestors, but the tools and weapons (and the brains) of those early humans became increasingly sophisticated as time went on. That gave the tool makers an increased ability to

Ecology, Ecosystems, and Agroecosystems

defend themselves and to become more powerful predators. With these tools, they would have been able to carve out new niches for themselves. This introduced a new dynamic into fitting one's environment – one that Mona had never dealt with. By dint of intelligence and dexterity, these species were adapting not just by genetic mutation and natural selection, i.e., evolution, but by using technologies that they invented. Learning how to use fire to warm themselves and cook food provided another highly significant technological advancement. It seems only the human brain is nimble enough to realize fire is not just to be feared. Some suggest that control of fire is the "invention" that made us ultimately human, and that milestone was passed at least 1.5 mya, when *Homo erectus* was a widely spread hominid. With tools and fire and bigger, smarter brains, humans became Mona's version of the terrible twos. As their technological prowess broadened, humans were beginning to express some independence from Mona and to assert themselves.

Then the wise humans – that's what *Homo sapiens* means – appeared. The first members of our species presumably inherited the technologies of their forebears, and then they laid language on top of that. Evolution provided *Homo sapiens* vocal mechanisms to produce distinctive sounds and brains that could interpret those sounds to mean objects, actions, and ideas. Now we could talk – and talk back. We were becoming teenagers.

Then the path of technology broadened again some 12,000 years ago, at a time when human numbers were only 4 or 5 million. The new technology was agriculture – teasing food out of prepared ground. Autobiologically, we can reason that the new way of earning a living would have led to the following sequence: Some proto-farmers became good

The Gyroscope of Life

enough (or industrious enough) to produce food for their families and have some left over for barter. Those surpluses freed up others to congregate into villages/towns/cities and become metallurgists, artists, poets, philosophers, physicians, inventors, investors, etc. By such reckoning, early agrarians were instrumental in the birth of civilization, making possible cities populated by non-farmers.

An ongoing debate among anthropologists concerns the relative merits of making one's living as a hunter-gatherer or a farmer. Some recent reports suggest tribes of hunter-gatherers in both Africa and Oceania have more leisure time than those who farm in the same places. Some who study cultural life styles have dubbed hunter-gatherers as "the original affluent society". That idyllic analysis is in sharp contrast to the belief of others that hunter-gatherers have always lived hardscrabble lives: perpetually on the brink of starvation, with shorter life spans, lowered fertility, and higher rates of infant mortality relative to agrarians. That's much more like the notion that hunter-gatherers live under Mona's thumb and fall victim regularly to the carrying capacity corollary of HR#1. A fundamentalist take on this argument comes down more on the hunter-gatherers' side – at least if they are foraging in a Garden-of-Eden-like place. When God sent Adam out of the Garden for eating the forbidden fruit, He further condemned him to farming: "…cursed is the ground because of you; through painful toil you will eat food from it all the days of your life. It will produce thorns and thistles for you, and you will eat the plants of the field. By the sweat of your brow you will eat your food…" (Genesis 3:15-17).

Whatever it meant for their leisure time, as farmers increased humanity's larders, they raised their locale's carrying capacities. Heretofore, Mona's carrying capacities for humans

Ecology, Ecosystems, and Agroecosystems

had been based on hunting and gathering – not on domesticating and cultivating plants and animals. As more hunter-gathers became more agrarian, they simultaneously became more citified, social, and entrepreneurial, and Mona lost more control. (Our evolving social institutions were based on economy – not ecology. No one understood then what Mona's House Rules were. Hell, most still don't.) Humans had something Mona had never dealt with – cleverness. At least none of her other children became intelligent on the scale that we have. Our technology is one of the surest signs of that inventiveness. The wheel was invented about 5,500 years ago; now we didn't even have to walk down the technology road.

(Just for the record, humans are not the only farmers. Leaf cutter ants, some termites, and some beetles cultivate fungi as food. Farmer ants include several ant species that are rather like dairy farmers, herding and milking aphids for their nectar-like secretions. Damselfish culture and protect algae that they then eat. These sometimes mutually beneficial arrangements developed by co-evolution of the farmer organism and the farmed organism. These relationships result from natural selection – not from cagey minds.)

Then, some 250 years ago (with still less than 1 billion of us on the globe), the Industrial Revolution essentially turned the road of technology into a superhighway. With rapid technological advances in food production, manufacture, medicine, transportation, communication, etc., the number of travelers on this superhighway increased in just a quarter-millennium by more than seven fold.

There is no doubt that our technology can support today's 7.7 billion humans. We are living proof, as it were. But one doesn't have to drill down too deeply into the numbers to

The Gyroscope of Life

see that many of us are living on the edge, with close to 1 billion going to bed hungry every night. And, if you look at the first sentence again, you will see that I say technology can "support" 7.7 billion. I did not say "sustain". There are very serious concerns about the sustainability – the "forever-ness" – of many of the technologies we employ, agricultural technologies included.

Modern agriculture is unsustainable by design, but not by intent:

Agroecosystem pretty much explains itself, as with ritualistic funerary cannibalism. Agroecosystems are the organisms and abiotic environment that interact in a human-managed agricultural setting. Those of us who study agroecosystems call ourselves agroecologists. (Agro- is from the Greek, of course, for a cultivated field.) For many years, I taught undergraduate and graduate courses in agroecology.

Real ecologists tend to look a bit askance at us agroecologists – like the stereotype of city folks looking down their noses at farmers. Many people have no idea that a farmer pays more for a big green tractor than they do for a big black SUV. They have little appreciation for the fact that modern farming is a high-tech business, with many famers being business-savvy college graduates. But real ecologists – even if they cannot agree among themselves what their field is all about – might have some grounds for smirking a bit when we agroecologists talk about farms as "agroecosystems". (Many ecologists have equal disdain for "urban ecosystems", by the way.)

If we boil it down, agriculture is rather antithetical to ecology. The nature – even the goal – of much of agriculture

Ecology, Ecosystems, and Agroecosystems

is to obliterate the flora and fauna of natural ecosystems and then to establish a non-native, domesticated species. Instead of adhering to the no-external-inputs dictum that characterizes natural ecosystems, modern agriculture relies heavily on importing exhaustible resources. Fertilizer must be shipped in and applied to fields. In many places, irrigation water is being pumped out of the ground at rates far exceeding Mona's ability to replenish it. Plows, which are commonly used worldwide, expose soil to losses at rates that exceed Mona's ability to make new soil. But we agroecologists still would like to describe our fields as agroecosystems. Maybe that agro- in front of those words is something of a fig leaf. (Back up in the hollers, they would say at this point, "Now you've stopped preachin' and gone to meddlin', preacher!")

Agriculture did not set out to be environmentally and ecologically unfriendly. The first farmers were barely beyond being hunter-gatherers. They probably had minimal impact on the ecosystems in which they lived, and subsistence farmers still do. But as our ancestors got more adept at turning fields into cornucopias, and as plants and animals became more and more tractable during domestication, agriculture grew in scale and intensity. Somewhere along the way, agriculture crossed a threshold; farming became a profession, a way of earning a living, not just a way of surviving and sustaining one's own family. Of course, that had tremendous benefits for society. While some could specialize in farming, others could focus on music, medicine, and making municipalities. But...

As the scale and intensity of agriculture grew, so did its potential to disrupt and destroy ecosystems. Domestication of draft animals, as romantic or positive as that might sound, was probably not an environmentally friendly step, because

The Gyroscope of Life

it increased our ability to farm even larger expanses. But, it was the Industrial Revolution that cast the doors wide open to large-scale disruptions, even obliterations, of ecosystems in the name of creating agroecosystems. Yes, most of us in the US and western Europe eat well today, and we can thank the 2% of us who are the farmers and ranchers providing all that food. But, all of that bounty comes at some cost to the environment. I believe the handwriting is on the wall that the current way we do agriculture is not sustainable. We have been weighed in the balance and found wanting.

Why modern agricultural methods are unsustainable:

Something of a thought experiment – an exercise in logic – will suggest how modern agricultural systems inevitably deplete finite resources and, in the process, make themselves unsustainable. Remember: sustainable suggests something has a quasi-infinite lifetime. Modern agroecosystems extensively use external, exhaustible inputs. That is in total contrast to the model of sustainable, no-external-inputs, perpetually productive ecosystems. The external inputs for agriculture as practiced today include the raw materials needed to manufacture tractors, plows, and other machinery and equipment. Then energy – almost always from finite fossil fuels – is used to make and run equipment, mine or make fertilizer, haul fertilizer, and haul products to the market. Then there are the nutrients that must be regularly imported and applied to fields.

Looking at the lifecycle of plant nutrients used in agriculture provides some of the clearest evidence for the unsustainability of modern practices. Typical farms are built on soils

Ecology, Ecosystems, and Agroecosystems

that once supported highly productive, nutrient-conserving, sustainable ecosystems. The lush prairies on the Great Plains, with their dark, rich soils, are just one example. However, the agroecosystems that replaced those natural perpetual motion machines must now be regularly replenished with nutrients – especially phosphate, potash (potassium), and nitrogenous fertilizers. In the US, phosphate fertilizer typically comes from deposits in Florida and Idaho. (In a geological oddity, over two-thirds of the world's phosphate reserves are found in Morocco.) Potash fertilizer for US farmers typically comes from deposits in Canada, which is home to almost one-quarter of the world's potash reserves. A third key nutrient, nitrogen, is commonly pulled out of the atmosphere and converted – at considerable fossil-fuel expense – into forms that plants can use. Wherever they come from, these nutrients must be transported – using fossil fuels – to farms and onto the fields.

Some portion of the imported nutrients will be captured by crops and boost yields, and that's good. But, then, they are hauled away when the harvest is sold. And whenever it rains, some of those applied nutrients will likely wash into streams or be carried deeper into the soil and out of the reach of plants' roots. In addition, nitrogen fertilizers can be volatilized back into the air. So, nutrients must be added yearly. The continual requirement for external, exhaustible nutrient inputs tells us the system is not sustainable, an ecological kiss of death. Modern high-input, or conventional, agriculture cannot produce food, feed, fiber, and fuel *ad infinitum*. They have no Gyroscope that can be kept spinning with just sunlight and water.

Some are making dire predictions for when we will run out of key inputs for agriculture. At current rates of use, the

The Gyroscope of Life

world's potash reserves will be depleted in about 600 years. For phosphate reserves, that would happen in about 260 years. That's just seven to 30 human generations.

But this sky-is-falling motif can and should be softened by noting that "reserve" is an economic, not a geologic, concept. The reserve is the currently known locations and amounts of a resource that can be economically extracted, processed, and delivered to users. As long as suppliers can extract and sell the resource at a profit, the reserve will continue to be used. But, even as it is being used, the reserve can grow. That's a head scratcher, until we realize that a reserve is not a fixed, or finite, quantity. It can increase for some easily understood reasons. For example, new, readily-extracted deposits of a resource might be discovered, or new technologies might improve extraction and processing efficiencies and open up already known deposits that had been too low-grade or contaminated (and therefore not considered part of the reserve). In either case, the reserve would expand.

When regarding the resource and not just its reserves, we have a pretty good "cushion" for both phosphate and potash. Some agronomists and economists suggest that, at current rates of usage, we should have enough phosphate to last for 1,500 years. For potash, the can could be kicked down the road for 7,000 years. If such projections are right, maybe this is a moot point for now. But the mere fact that we are using these nonrenewable resources still labels these agricultural practices or systems as unsustainable.

Eutrophication; it's uglier than it sounds:

Relying on exhaustible, external inputs is not the only reason modern agroecosystems must be labeled unsustainable.

Ecology, Ecosystems, and Agroecosystems

Agriculture is unsustainable when it damages natural ecosystems. (This would be in addition to the ecosystems that were obliterated in the act of creating agroecosystems.) Damage to aquatic ecosystems provide just one example. When fertilizers with nitrogen or phosphate wash into a body of water, they become pollutants. Understanding how something essential for productive agroecosystems can be bad for aquatic ecosystems requires some knowledge of a phenomenon called eutrophication.

Some bodies of water – Lake Tahoe is a beautiful example – are gin clear because they are essentially algae-free. Algae cannot grow because nitrogen and phosphate, which are essential nutrients for algal growth, are in such short supply. But, if nitrogen and phosphate are introduced into one of these clear bodies of water, algae will begin to grow. The population of algae can potentially explode. (It's call an algal "bloom", but it's not pretty.) The water can begin to look more like pea soup. And, because the water is so thick with algae, sunlight cannot penetrate deeply. Algae in the dark begin to die.

As the algae die, bacteria begin to break them down. It's all part of the wisdom displayed by ecosystems – turning over nutrients so they can be reused. But the bacteria use oxygen dissolved in the water to carry out the decomposition of the dead algae. Before long, all of the oxygen in the water has been used up. That creates a nightmare for the oxygen-requiring organisms living there. Fish, crawdads, shrimp, lobsters, insect larvae, mollusks, salamanders, and plants are asphyxiated. Major portions of the Chesapeake Bay become oxygen-starved each summer. Nothing that requires oxygen can live in the "dead zone". Something similar happens where the Mississippi River runs into the Gulf of Mexico,

The Gyroscope of Life

but the scale of the ugliness is even larger. A dead zone the size of New Jersey extends annually out into the Gulf at depths of about 30 to 60 feet. The oxygen-starved zone is at its peak during the summer, when nutrient levels and bacterial activity are highest.

In some parts of the Chesapeake Bay watershed, restrictions have been put on fertilizer use. While initially resistant, many farmers are learning that it makes both economic and environmental sense to adopt new fertilization practices, and the Chesapeake Bay estuary, which used to be one of the world's most productive ecosystems, is now enjoying something of a comeback. The problem of the dead zone in the Gulf of Mexico has proven to be less tractable. There are possible solutions, but the politics of institutionalizing them are daunting.

(I have not discussed the extensive use of pesticides in modern agriculture. They are typically employed to counter specific pests, but they are not smart bombs; they often cause collateral damage in down-wind and down-stream ecosystems. This does not fit the definition of sustainable. It reminds us of Aldo Leopold's observation: "A thing is right when it tends to preserve the integrity, stability, and beauty of the biotic community. It is wrong when it tends otherwise.")

A restatement of the problem

So, we come back to the assertion that modern, large-scale agriculture is inherently not sustainable. Even if we can make it harmless to bodies of water and other ecosystems, high-input agriculture is not sustainable because it relies so heavily on external, exhaustible inputs. One of these days, we are go-

Ecology, Ecosystems, and Agroecosystems

ing to deplete readily available phosphate and potash deposits. One of these days, we are going to run out of fossil fuels. Add to those inevitabilities the fact that farmland acreages have been declining across the globe for several years. Throw in the statistic that soil erosion is making some agricultural land less and less productive even as more nutrients are being used. Some suggest that western-style agriculture is on a treadmill; productivity and economic viability can be maintained only with more inputs and bigger farms. Some sort of agro-Darwinism seems to have taken over.

Modern farmers rise to the challenge of feeding the world and generally have great respect for the land. Many now emphasize stewardship of the land and of the environment. Unfortunately, that's not going to matter when potash and phosphate resources run out. The 2nd Law of Thermodynamics suggests we will never be able to re-collect the potash- and phosphorus-rich deposits that have been dispersed into entropic oblivion. No salmon are swimming back, dying, and replenishing the nutrients in agroecosystems.

Alternatives to unsustainable agricultural practices have a steep hill to climb:

Some tout "organic" agriculture as a retro-solution to today's exhaustive and pollutive style of agriculture. I put organic in quotes, because the word means different things. For our purposes, let's let it nebulously mean non-exhaustive and non-pollutive. Some farmers are returning to practices that characterized pre-20th-century agriculture, where external inputs are minimized. Such practices include putting animal manure on fields to provide nutrients needed by crops. In a well-run, crop-plus-livestock agroecosystem, nutrients

The Gyroscope of Life

can be cycled internally – just as in the ecosystems where buffalo roamed and the deer and the antelope played. A farm with both crops and livestock could mimic the prairie grassland's nutrient loops. At least it could work when considered as a thought experiment. But, if crops or animals are sold off this fanciful farm, the experiment fails. As they are hauled away, those sold products become a conduit for loss of nutrients. Selling eggs, milk, cheese, flowers, etc. would cause additional losses from the loop; small losses perhaps, but even "insignificant" losses would become significant over time. And sustainability is all about what happens over time. Nitrogen fertilizer can be supplied organically/non-exhaustively by planting legumes. But, there are no organic/sustainable methods for generating the other nutrients that go trundling down the road to market.

Here is another major rub in scenarios to organically feed 7.7 billion people. In the last few years, several peer-reviewed scientific articles have examined in meta-analysis style the results of experiments where conventional methods and organic methods have been compared side by side. In meta-analysis studies, authors bring together data from hundreds of studies, analyze them as one meta-data set, and then draw conclusions. My "auto" analysis of those meta-analyses suggests that organic systems typically have 10 to 30% lower yields than conventional systems, although some organic systems were more productive. With one in nine of us (globally) going to bed hungry most nights, it seems we might put even more of us at risk by stepping off of the conventional treadmill.

Of course, many parts of the world are not on the conventional treadmill. Many small landholders do a good job of taking advantage of niche, internal cycling of nutrients,

Ecology, Ecosystems, and Agroecosystems

and other non-exhaustive strategies. Unfortunately, such more-nearly-natural agroecosystems often do not scale-up well. They are typically used in labor-intensive, subsistence farming and can be difficult to adapt to a commercial level. Whenever we consider commercial-scale farming, we generally introduce the need for external energy and nutrient inputs.

Some tout going to the opposite end of the technology spectrum to feed the world. Their blue sky (or pie in the sky) thinking foresees a time when agriculture produces more food per acre. They prophesy that genetic engineering of crops and livestock – genetically modified organisms (GMOs) – will cause those domesticated beings to greatly step up their output. These prophets of profits, may be proven right. I am skeptical. Mona has been working with terrestrial ecosystems for almost a half-billion years. Her best ecosystems are about 2% efficient in converting sunlight into living matter. When we compare the amount of energy in a harvested crop to the amount of sunlight energy that fell on the land that grew it, the best agroecosystems in temperate zones can achieve only about one-quarter of what Mona can. It would take some pretty far out GMO solutions to break through what appears to be a significant yield barrier. We're not talking about the equivalent of a 4-minute mile. It would be more like a 2-minute mile. And, even if plants can be jiggered to be more productive, no one is suggesting they can be weaned of their need for potash and phosphate.

(I do not feel qualified to address the many health and environmental concerns raised with the increasing use of GMOs. I will venture, though, that we probably understand as much about potential harms from GMOs as we do about managing ecosystems.)

The Gyroscope of Life

Necessity is the master of intention:

Agricultural systems designed to feed 7.7 billion inevitably require external inputs and are therefore unsustainable – not the perpetual motion machines that natural ecosystems are. Only large-scale agriculture can feed the 98% of Americans and Western Europeans who aren't farmers. (Globally, the percentage of non-farmers is about 75.) So, 2% of us (25% globally) provide the rest of us food, feed, and fiber. We are, in essence, paying farmers to convert natural ecosystems into agroecosystems to provision us cheaply and abundantly but at a high price environmentally. As Pogo said, "We have met the enemy, and he is us." "Us" is not the farmers. It is all of us, all 7.7 billion of us, and especially those of us who are such conspicuous and copious consumers.

We currently use about 3.5 billion acres globally for annually planted crops – crops like wheat, rice, corn, and soybean. The math is portentous – less than half an acre (about half a football field) feeds each mouth. We make that work (more or less) with high-input, minimal-hand-labor agroecosystems. (About half of the food grown globally today is fertilized with off-farm inputs.) Could we make agriculture sustainable if all of us worked, lived, and ate on farms? Maybe, but then who would be the doctors, manufacturers, police, poets, philosophers, etc. It's not just farmers; civilization itself is on the treadmill.

We all share one home – Earth. It is our little space ship. The "little space ship" quote that opened this chapter was from a speech that Adlai Stevenson II, two-time candidate for US President, gave to the UN Economic and Social Council 5 days before his death in 1965. Here is my heartfelt "Amen" to his exhortation that we find solutions to the

Ecology, Ecosystems, and Agroecosystems

life-threatening problems we are visiting upon ourselves and our home. I am sanguine that there are solutions. We just need to move toward them before Mona moves toward her hook...

Further Reading

I have resisted providing citations and footnotes throughout the book. However, I know some readers like to have information on relevant resources. Here are a few titles that bear on topics I have covered (but not nearly so well as these authors).

Bill Bryson (2003) *A Short History of Nearly Everything*. A fascinating, very readable take on many topics to include some of those I have discussed.

Rachel Carson (1962) *Silent Spring*. Carson's scathing report on the state of the environment in the mid-20th century and her momentous call to arms. Meticulously documented and damning.

Richard Dawkins (1976) *The Selfish Gene*. This highly influential book offers a gene-centered view of evolution. It makes a case for genes being more important biologically than the organisms that carry them. Altruistic behavior, which is seen in many species, is turned inside out and deemed a self-serving way to favor gene transmission.

Paul R. Ehrlich (1968) *The Population Bomb*. Ehrlich sounds an alarm about overpopulation – perhaps a bit too shrilly. His arguments are built primarily around the tenuousness of food supplies. He argues strongly for steps to reverse population growth.

Yuval Noah Harari (2014) *Sapiens: A Brief History of Humankind*. A fascinating take by a historian on how things may have unfolded for our species from prehistory into the present. Harari takes a rather dim view of agriculture.

Jack R. Harlan (1992) *Crops and Man (2nd edition)*. Harlan, an agronomist, provides a very readable story on the origins of agriculture and the domestication of crops.

Stephen W. Hawking (1988) *A Brief History of Time: From the Big Bang to Black Holes*. Written by one of the 21st century's most brilliant minds (trapped in a paralyzed body), this book makes understandable to those not mathematically inclined some mind-boggling details about cosmology and astrophysics.

Hope Jahren (2016) *Lab Girl*. A very informative, funny, and passionate account by a very bright plant and soil scientist. This is the book that made me think I might want to write about biology to non-biologists.

Aldo Leopold (1949) *Sand County Almanac*. This short book of memoirs and essays is widely recognized as a groundbreaker and a landmark in American ecology and environmentalism. Leopold's prose approaches that of Thoreau in capturing the beauty, and the starkness, of nature.

James Lovelock (1979) *Gaia: A New Look at Life on Earth*. Lovelock makes an excellent case for why all of life on Earth might be considered a single, self-regulating super-organism.

Desmond Morris (1967) *The Naked Ape: A Zoologist's Study of the Human Animal*. Morris champions the notion that human behaviors are evolutionary products, especially as related to sexuality. A provocative read.

E.O. Wilson (2002) *The Future of Life*. Wilson is the world's expert on ants and ant societies, one of the world's most respected ecologists, and a founder of the field of sociobiology. In this book, he decries accelerating rates of extinction of life forms. He identifies causes

The Gyroscope of Life

and consequences and suggests life-saving remedies.

Carl Zimmer (2014) *The Tangled Bank: An Introduction to Evolution (2nd Edition)*. A textbook written for general readers. Full of facts and interesting stories. Discusses the history of evolution, evolutionary biologists' tools of the trade, how evolution works, and much more.

Author Bio

David Parrish is proud to have been born and bred in Appalachia, and he counts himself fortunate to have worked and played in those same fascinating hills throughout his professional life. After earning a doctorate in plant science at Cornell, he joined the faculty of Virginia Tech's College of Agriculture and Life Sciences, where he was recognized with awards for excellence in teaching and advising. Courses he taught include crop science, crop ecology, and environmental science.

Parrish has co-authored a textbook on plant science, more than 50 research articles, and several scientific reviews and book chapters. His research interests spanned plant growth regulation, seed physiology, sustainable cropping systems, and biological sources for renewable energy.

In retirement, Parrish and his wife of more than 50 years remain in the mountains of southwestern Virginia on the 30 acres they call Hole in the Woods. He continues to be an avid gardener and outdoorsman, particularly in chasing catch-and-release smallmouth bass on the New River.

www.ingramcontent.com/pod-product-compliance
Lightning Source LLC
Chambersburg PA
CBHW070913030426
42336CB00014BA/2390